Sustaining FORESTS

A Development Strategy

THE WORLD BANK
Washington, D.C.

1818 H Street, NW
Washington, DC 20433
Telephone 202-473-1000
Internet www.worldbank.org
E-mail feedback@worldbank.org

04 05 06 07 4 3 2 1

Library of Congress Cataloging-in-Publication Data

Sustaining forests : a development strategy.
p. cm.
Includes bibliographical references (p.)
ISBN 0-8213-5755-7
1. Forests and forestry—Economic aspects. 2. Forest policy. 3. World Bank. 4. Sustainable development. I. World Bank.
SD393.S87 2004
333.75—dc22 2004043984

Contents

BOXES

FIGURE

TABLE

Appendixes

Note: These appendixes are included on the CD-ROM that accompanies this book.

BOXES

FIGURES

TABLES

Preface

The World Bank Group consists of closely associated but legally and financially distinct institutions that support development in low- and middle-income client countries.

The International Bank for Reconstruction and Development (IBRD) and the International Development Association have a mandate to lend to sovereign governments. They often are referred to as the World Bank. Through its ongoing dialogue with client governments on a wide range of issues and its financial support to the public sector, the World Bank is in a position to support government policy on a wide range of issues.

The International Finance Corporation (IFC) promotes sustainable private sector investment to enhance economic growth and improve the lives of the poor. It operates principally through direct or indirect support of private sector projects.

The Multilateral Investment Guarantee Agency (MIGA) provides guarantees against certain noncommercial risks (primarily political risk insurance) to foreign investors for qualified investments in developing countries.

The International Centre for Settlement of Investment Disputes provides facilities for the settlement—by conciliation or arbitration—of investment disputes between foreign investors and their host countries.

The five World Bank Group institutions are aligned to the core mission of poverty reduction; therefore, the overall vision, strategic framework, and objectives of this Forest Strategy are shared by the entire World Bank Group. The specific implementation tasks identified in the Forest Strategy document are addressed to the World Bank. IFC and MIGA will implement the Forestry Strategy through their ongoing emphasis on financing/insuring private sector investments that improve forest management and sustainable outcomes. As for IFC's forestry policy, it should be noted that all of IFC's environmental and social safeguard policies are the subject of a comprehensive review by IFC's Office of the Compliance Advisor and Ombudsman. In addition, any future safeguard policy formulation for the forestry sector needs to be congruent with the findings of that review and consistent with IFC's need to provide clear policy guidance for private sector investments. For these reasons, the World Bank's approach to the Forestry Policy described in chapter 2 of this strategy may not apply in its entirety to IFC.

This book is accompanied by a CD containing background materials on how the World Bank's Forests Strategy was developed, including the stakeholder consultative process, as well as information on the role of forests in poverty reduction, economic development, and the provision of environmental services that helped to shape the strategy. World Bank safeguard policies relevant to forests and a short video highlighting the strategy's objectives are also included in the CD.

Acronyms and Abbreviations

AFR	Africa Region
CAS	Country Assistance Strategy
CBD	Convention on Biological Diversity
CDF	Comprehensive Development Framework
CDM	Clean Development Mechanism
CEO	Chief executive officer
CEPF	Critical Ecosystem Partnership Fund
CFM	Collaborative forest management
CI	Conservation International
CIFOR	Center for International Forest Research
CGIAR	Consultative Group on International Agricultural Research
CODE	Committee on Development Effectiveness
CPF	Collaborative Partnership on Forests
CTF	Conservation Trust Fund
EAG	External advisory group
EAP	East Asia and Pacific Region
ECA	Europe and Central Asia Region
ECOSOC	United Nations Economic and Social Council
ENFP	Enhanced national forest program
ESSD	Environmentally and Socially Sustainable Development Vice Presidency (or Network)
ESW	Economic and sector work
EU	European Union
FAO	United Nations Food and Agriculture Organization
FPIRS	Forest Policy Implementation Review and Strategy
FSC	Forest Stewardship Council
FSSP	Forest sector strategy paper
GDP	Gross domestic product
GEF	Global Environment Facility
GNP	Gross national product
ha	Hectare
HIPC	Heavily indebted poor countries
IBRD	International Bank for Reconstruction and Development
ICRAF	International Centre for Research in Agroforestry
IDA	International Development Association
IFC	International Finance Corporation
IFF	Intergovernmental Forum on Forests
IIED	International Institute for Environment and Development
IMF	International Monetary Fund
IPF	Intergovernmental Panel on Forests
ITFF	Interagency Task Force on Forests
ITTO	International Tropical Timber Organization
IUCN	The World Conservation Union

IUFRO	International Union of Forest Research Organizations
JFM	Joint forest management
LAC	Latin America and Caribbean Region
MDGs	Millennium Development Goals
MIGA	Multilateral Investment Guarantee Agency
MENA	Middle East and North Africa Region
NGO	Nongovernmental organization
NFP	National forest program
NTFPs	Nontimber forest products
OD	Operational Directive
OECD	Organisation for Economic Co-operation and Development
OED	Operations Evaluation Department
OP	Operational Policy
OPCS	Operations Policy and Country Services
PCF	Prototype Carbon Fund
PROFOR	Program on Forests
PRSP	Poverty reduction strategy paper
SAR	South Asia Region
SFM	Sustainable forest management
TAG	Technical advisory group
TNC	The Nature Conservancy
UN	United Nations
UNDP	United Nations Development Programme
UNEP	United Nations Environment Programme
UNFF	United Nations Forum on Forests
UNFCCC	United Nations Framework Convention on Climate Change
UNGASS	United Nations General Assembly Special Session for Review and Appraisal of Agenda 21
WBI	World Bank Institute
WRI	World Resources Institute
WWF	World Wide Fund for Nature/World Wildlife Fund

Executive Summary

THE CHALLENGES

More than 150 heads of state have declared that halving extreme poverty by 2015 is central to the United Nations' Millennium Development Goals (MDGs).[1] These goals include increasing school enrollment, reducing child and maternal mortality, expanding health services, eliminating gender disparities, and improving environmental management for sustainable development. The World Bank supports these goals by emphasizing the social and structural dimensions of development, focusing its efforts, increasing selectivity, and emphasizing partnerships and transparency. As a result, the Bank is pursuing global and corporate advocacy priorities and areas of core competencies.

A Forest Strategy for the Bank that can make an effective contribution to poverty reduction and environmental management is central to achieving the MDGs. Forest resources directly contribute to the livelihoods of 90 percent of the 1.2 billion people living in extreme poverty and indirectly support the natural environment that nourishes agriculture and the food supplies of nearly half the population of the developing world. Forests also are central to growth in many developing countries through trade and industrial development. However, mismanagement of this resource has cost governments revenues that exceed World Bank lending to these countries. Illegal logging results in additional losses of at least US$10 billion to US$15 billion per year of forest resources from public lands. If captured by governments, these losses could support expenditures in education and health that will exceed current development assistance to these sectors.

Forests also are central to maintaining the environmental commons. Nearly 90 percent of terrestrial biodiversity is found in the world's forests, with a disproportionate share in the forests of developing countries. Most of the carbon emissions of developing countries come from deforestation, which accounts for between 10 and 30 percent of global carbon emissions. Unfortunately, the lack of markets for the national and global environmental services offered by forests has contributed to high rates of deforestation in developing countries. Growing forests are a valuable resource not just for their timber and biodiversity values but also for their prospective value if a global market emerges for the sequestering of carbon from forests.

WORLD BANK'S PERFORMANCE IN THE FOREST SECTOR

The Bank's performance in the forest sector over the past decade has been unsatisfactory. The World Bank's 1991 Forest Strategy and 1993 Forest Policy focused largely on environmental issues and protecting tropical moist forests. They reflected rising international concern about the rate of tropical deforestation and strongly emphasized the need to preserve intact forest areas. While the 1991 Strategy recognized the role that forests could play in poverty reduction and the importance of policy reforms in containing deforestation, its hallmark was a strong commitment not to finance commercial logging in primary tropical moist forests.

In practice, this emphasis on safeguarding forests has meant that little attention was paid to the active management of natural forests in the tropics and therefore to the poverty-reduction potential of forests. The 1993 Policy led to a generally passive "do no harm" stance on natural forests in the tropics. Interventions designed to more proactively improve economic and environmental management of those forests often were seen as too costly and risky. What the World Bank's Operations Evaluation Department (OED) has termed "the chilling effect" permeated operations in the Bank Group, including the International Finance Corporation (IFC) and Multilateral Investment Guarantee Agency (MIGA) (Lele and others 2000f).

Furthermore, the 1991 Forest Strategy did not clearly define implementation mechanisms. As a consequence, countries rich in forest resources have not received World Bank funding, nor have they benefited from strong research and analytical sector work (in the Bank, termed economic and sector work, or ESW) or engaged in a strong dialogue on Forest Policy and Strategy. The bulk of the US$3.7 billion lent by the World Bank for forests during the 1990s went to China, India, and Eastern Europe. The OED review concluded that the Bank had been "irrelevant" in slowing deforestation despite its commitment to this objective in its 1991 Forest Strategy. Therefore, the World Bank needs to modify its strategy, expand its policy to explicitly include all forest areas, and refocus the strategy on poverty reduction and economic management, including good governance.

Moreover, it is now acknowledged that the impacts on forests and forest-dependent peoples of what the Bank does in support of policy reforms and investments *outside* the forest sector are equal to, or even greater than, its forest sector activities. Nonforest interventions, such as rural development and infrastructure programs and projects and economic adjustment measures, must be carefully formulated to take account of their influence on forest outcomes.

ELEMENTS OF THE STRATEGY

The Bank's Comprehensive Development Framework (CDF) requires the institution to retain a broad diagnostic and analytical capability but be selective in its direct assistance, seeking its comparative advantage and working with clients and partners. Through this approach, it is expected that the Bank will maintain a high degree of economic and technical work while consulting widely with partners and stakeholders.

Building the Bank's new Forest Strategy has followed this model of analysis and consultations. A two-year process of analysis and consultation gathered input from development partners and stakeholders around the world, including governments, civil society in developing and developed countries, industry, forest-dependent peoples, and partner donor agencies and UN bodies (appendix 7). The process also included broad consultations with Bank task managers and other staff.

These consultations were supported by extensive analytical, technical, and economic studies, some commissioned by the World Bank and others done in parallel by independent institutions and nongovernmental organizations (NGOs) on a wide range of subjects (see appendix 8). Among other things, these studies emphasized the importance of addressing governance and quantified the costs of poor forest management and illegal activities for several countries. The studies also confirmed the close link between the livelihoods of the poor and forests and helped to refute the largely false notion that the poor are the cause of deforestation in developing countries. The studies also identified strategic approaches to forest issues, including participation and partnerships.

From this process, the revised Forest Strategy has been built on three equally important and interdependent pillars:

- Harnessing the potential of forests to reduce poverty
- Integrating forests in sustainable economic development
- Protecting vital local and global environmental services and values

Addressing these three aspects together makes a Forest Strategy complex and multifaceted. It is not only about growing or protecting trees but also involves a complex interaction of policy, institutions, and incentives. A narrow perspective on forestry—even sustainable forestry—is insufficient. To be effective, the strategy demands a multisectoral approach that addresses cross-sectoral issues and takes into account the impacts of activities, policies, and practices outside the sector on forests and people who depend on forests for their livelihoods.

Within these three pillars of engagement, the Bank must be selective in the activities it supports. In broad terms, the Bank will focus on economic policy and rural strategies that embrace both conservation of vital environmental services and sustainable use. It will provide institutional and policy support for community and joint forest management, governance and control of illegal activities, building markets, and financial instruments in support of private investment in sustainable forest conservation and management. It will emphasize the development of new markets and marketing arrangements for the full range of goods and environmental services available from well-managed forests. For the IFC and the MIGA, the major focus will be to support private investments in sustainable forest management (SFM), conservation, and rural forest industries.

Harnessing the Potential of Forests to Reduce Poverty

Implementation of the revised Forest Strategy will make a significant contribution to meeting the goals of the Bank's revised Rural Development Strategy. This strategy refocuses the rural development process to concentrate on improving the well-being of rural people and reducing poverty in the widest sense. The latter entails much more than increasing the average income of rural people. It envisions improving the quality of rural life. The underlying concept for this strategy is a developing world in which

- Rural residents enjoy a standard of living and a quality of life that are not significantly below that available to urban residents;
- Rural communities offer equitable economic opportunities for all their residents regardless of income, status, or gender;
- Rural communities become vibrant, sustainable, and attractive places to live and work;
- Rural areas contribute to national development and the overall economy and are dynamically linked to urban areas;
- Rural areas adapt to ongoing economic, social, cultural, and technological change.

Analytical studies and field experience show that forest outcomes are crucial for poverty reduction in many of the Bank's client countries. This result is true in countries with large forest endowments as well as in those with limited forests, although the nature and urgency of the challenge may vary. If forest issues are not fully incorporated in a broad assistance strategy, the broader goals of poverty reduction will not be met.

Effectively addressing the poverty issues related to forests is not straightforward. Experience has shown that remedial strategies can generate internal conflicts. Assistance should be provided judiciously to those dependent on or who live near forests so that they may develop their abilities to service the forest products market. If this is not done correctly, it could increase competition for the forests, exclude access to the poorest of the poor to essential forest products, and disrupt communal systems of management by groups that traditionally have relied on common property forest resources for meeting essential fuelwood, grazing, and other needs. The Bank will need to rely on its partners—particularly civil society—and on pilot operations supported by others to demonstrate feasible approaches that can then be scaled up to make a significant contribution to social, environmental, and economic objectives.

Within this framework, the Bank's strategy will focus on creating economic opportunity, empowerment, and security for rural people, especially the poor and indigenous groups. The main instrument will be through policy and institutional strengthening to ensure that the rural poor are able to manage their natural resources, especially forests, for their own benefit. It will assist governments in building the capacity to support and regulate community use of forests and plantations. Special attention will be directed to the welfare of some 60 million indigenous tribal peoples living in the rain forests of West Africa, Latin America, and Southeast Asia who have high levels of dependence on forest resources. Impacts will be reflected in strengthened tenure rights, improved food security, and spiritual welfare among the poorest of the poor through assured access to essential forest products on which they depend primarily for subsistence.

Equally important will be the Bank's support to collaborative forest management systems that are emerging in India and many other parts of the world and that have proved to be both beneficial and sustainable when managed correctly. Contemporary approaches to Protected Areas include the concepts of the World Conservation Union (IUCN) Category VI which cover "areas containing predominantly natural systems managed to ensure long-term protection and maintenance of biological diversity while providing at the same time a sustainable flow of natural products and services to meet community needs." Using this broader definition suggests that community-managed forests, for example, those in India, also should be recognized for their major contribution to the preservation of biodiversity, carbon sequestration, and other environmental services. In support of forest operations, IFC and MIGA have important roles in continuing to support private investments in SFM, conservation, and rural forest industries. In collaboration with its client countries and partners, the Bank's primary roles will be

- To work with client countries to strengthen policy, institutional, and legal frameworks to ensure the rights of people and communities living in and near forest areas;
- To ensure that women, the poor, and other marginalized groups in society are able to take a more active role in formulating and implementing rural forest policies and programs;
- To support the scaling up of collaborative and community forest management so that local people can manage their own resources, freely market forest products, and benefit from security of tenure;
- To work with local groups, NGOs, and other partners to integrate forest, agroforestry, and small enterprise activities in rural development strategies.

Integrating Forests in Sustainable Economic Development

The analytical work and consultations have demonstrated that forests are one of the developing countries' most mismanaged resources. The reasons are that forests are seriously undervalued, many of their environmental benefits do not enter markets, and poor governance has fueled illegal activities. In addition, the impact on forests of policy and investments in other sectors is not well understood or is disregarded. Although some level of deforestation is likely to continue even with strong economic management and governance, the rapid rates of deforestation in recent decades are largely a result of the spillover of poor policies in other sectors, including macroeconomic and trade policy, and lack of effective governance in the sector.

The Bank will need to focus on helping governments improve policy, economic management, and governance in the forest sector, including forest concessions policy and allocations. By supporting applied research and improving economic analysis, the Bank will assist governments to evaluate their policies and interventions and support reforms through adjustment credits and loans. It will also need to examine the nature of those adjustment operations more carefully, in terms of their potential impacts on forests. During the consultative phase in formulation of this strategy, a number of commentators noted that the proposed new forestry policy did not address potential forestry impacts of programs supported by Bank adjustment lending. Because the time frame for an update of the Bank's adjustment lending policy remained uncertain, some suggested that the Bank put in place a transparent set of procedures for systematically identifying significant forestry impacts associated with Bank adjustment operations, analyzing such impacts, and, if necessary, adopting and implementing appropriate mitigating measures.

In response to these concerns, the Bank has developed an approach to deal with development of a new adjustment lending Operational Policy (OP) in a timely manner and to address potential problems in the intervening period. This approach is outlined in chapter 2, in the section on "Integrating Forests in Sustainable Economic Development."

The Bank also will support government efforts to bring about socially, ecologically, and economically sound management of production forests. In this regard, the Bank also will encourage independent monitoring and certification of forest operations—an increasingly accepted approach to ensuring good forest management. Independent monitoring and certification will be additional to the Bank's regular implementation and safeguard procedures. It will help ensure that any direct Bank Group investments in production forests or indirect support through financial intermediaries or forest industries are contributing to improved forest management and more sustainable outcomes, including the protection of biodiversity and ecologically and culturally sensitive areas.

In supporting independent certification, the Bank will not endorse any one particular approach to certification. Outside the Bank there is an ongoing discussion on "mutual recognition" among various certification groups that is seeking to harmonize standards and approaches. However, in the absence of any broad stakeholder consensus on the acceptability of particular systems, the Bank has adopted a set of principles and criteria to assess the adequacy of different certification systems in relation to recognized standards of economically, environmentally, and socially sustainable development. These principles and criteria are discussed in chapter 2 of this report and have been incorporated as a screening tool in the OP. A key requirement is that the certification system incorporates reliable and independent assessment procedures.

Formal market-based certification systems are finding increasing use in situations where a significant proportion of forest production enters environmentally discriminating domestic or international markets. They are less relevant in situations where the bulk of production will go to nonenvironmentally discriminating domestic markets. In these situations, the Bank will work with borrowers to ensure independent assessment through an open process of review by the main participants and interest groups involved or affected by the forest use in question. These evaluations will be based on the standard and objectives of the operation that will be established with these groups at the outset.

The Bank has also worked with a number of partners to develop diagnostic tools based on a staged modus operandi that allows for periodic independent assessment of the progressive steps needed to put in place a sound legislative, institutional, and fiscal policy framework for improved forest management.

The Bank's primary roles in supporting the objectives of the second pillar will be as follows:

- To analyze and coordinate policies and projects to ensure a cross-sectoral approach to planning and implementation of SFM, conservation, and development. In this context, the Bank will support governments in making policy and institutional reforms to reduce the pressures on forests from activities and interventions in other sectors
- To support improved governance through reform of inappropriate timber concession and subsidy policies and the encouragement of multistakeholder involvement in the development and implementation of Forest Policy and practice
- To assist governments in containing illegal activities and corruption through improved forest laws, regulations, and enforcement
- To address finance, fiscal, and trade issues related to the forest sector and forest products to enable governments to capture a higher portion of forest revenues for sustainable social and economic development. The Bank will promote catalytic investments in the full range of goods and environmental services available from well-managed forests. These investments will be able to include sustainable timber harvesting and management, but only in areas outside critical forest conservation areas in situations that can be independently monitored through a system of independent verification or certification that meets nationally agreed and internationally acceptable standards

Protecting Vital Local and Global Environmental Services and Values

Implementation of the revised Forest Strategy will make a significant contribution to meeting the goals of the Bank's July 2001 Environment Strategy. The three pillars of the revised Forest Strategy correspond closely with the main objectives of the Environment Strategy: improving the quality of life, improving the quality of growth, and improving the quality of global commons. Both strategies recognize the importance of cross-sectional issues, mainstreaming policy dialogue, governance, designing interventions with selectivity, and fostering better cooperation with development partners. An area of particular concern to both strategies is the protection of the vital local and global environmental services provided by forest ecosystems.

Protected Areas have been the traditional method used to protect biodiversity and other key environmental services. At present, more than 600 million hectares (ha) of Protected Areas have been established in developing countries. However, while many of these areas are economically inaccessible, other areas are under increasing pressure from development and illegal activities, including logging and poaching. Many governments do not have the resources to effectively administer and protect these areas. In addition, other forests outside Protected Areas that are ecologically sensitive and rich in biodiversity are under increasing threat.

OED noted the inherent difficulty in protecting forests that are in high demand for a range of frequently mutually exclusive uses by competing groups within society. Although the Bank has generated significant increases in funding for biodiversity protection and related purposes, this contribution is dwarfed by the incursions into forests. The problem of invasive pressures is likely to worsen unless significant additional funds for protection can be made available from multiple sources, at highly concessional or on grant terms, or unless effective markets for the ecosystem values of forests can be developed.

Because of these difficulties, few countries in either the developing or the developed world have been either willing or able to devote more than

10 to 20 percent of their forest areas to strict conservation reserves. Given these constraints and the likely stresses associated with accelerating climate change and the continued growth in human populations and activity, it is clear that the future of forests and their dependent biodiversity and human populations is going to be influenced as much—and probably even more—by how forests outside strict Protected Areas are managed. Thus, improving forest management practices in production forests outside strict Protected Areas is an essential component of any strategy to protect vital local environmental services and values.

The Bank will assist in this area by developing and funding mechanisms to ensure that national parks receive the minimum funding needed for effective management and in building and facilitating new markets for forest ecosystem services. The increased revenues and incomes that national governments and local communities can gain from these environmental services can serve as a major incentive to sustain forests.

The Bank's primary role in implementing the third pillar will be

- To assist governments to proactively identify and conserve critical forest conservation areas in all forest types in all borrower countries;
- To assist governments to promote the wide-scale adoption of responsible forest management practices in production forest outside critical forest conservation areas;
- To develop options to build markets and finance for international public goods such as biodiversity and carbon;
- To assist governments to develop measures to mitigate and adapt to the anticipated impacts of climate change and reduce the vulnerability of the poorest people to its effects;
- To assist governments to design, implement, and finance national markets for local environmental services provided by forests;
- To assist governments to strengthen forest investments, policies, and institutions to ensure that indirect and cross-sectoral impacts of policy and investments on high conservation and protection areas are minimized;
- To ensure that Bank investments and programs in the forest sector and other sectors that could potentially harm protected forests and natural habitats are implemented according to the Bank's operational policies and safeguards.

THE IMPORTANCE OF COUNTRY OWNERSHIP

As will be elaborated upon throughout this paper, a major reason for adopting the three pillars of strategy outlined above is that they encompass the broader interests that borrower governments and other local stakeholders have in the management and use of their forests.

The success of the new approach will be completely dependent on the development of demand for Bank involvement in various ways from governments and other interest groups. It will also depend on their willingness to take a leadership role in the analytical and consultative work that underpins this approach, so that the results of this work genuinely develop country-level consensus and commitment to whatever lending or nonlending activities are proposed for that country. The partnerships and processes proposed in this strategy to implement the new approach have been chosen with a view to their compatibility with country ownership.

IMPLEMENTING THE STRATEGY: LARGE OBJECTIVES, MODEST BEGINNINGS

The importance of the forest sector to poverty reduction, sustainable development, and the maintenance of environmental services and values is recognized. The gap between performance and vision needs to be closed. The Bank's effectiveness in forests has been declining in some important regions of the world. As pointed out by OED, involvement in the forest sector is often seen as being high cost and risky. To increase engagement and demand on the part of some major client governments, the benefits of forest operations need to be demonstrated, the costs of doing business reduced, and the risks better managed.

The tasks involved in the three pillars are large, but the potential outcomes for forests, forest-dependent people, and countries with large forest resources are even larger (see page 13 below). The program for the Bank to reengage in the forest sector therefore begins modestly, recognizing that time and effort will be needed to build the institutional capacity and the expanded demand from borrowers for Bank involvement to a level that will be on a scale with the broad goals. In its early stages, it is based on analysis and consultative work, linked to Bank and national forest program (NFP) priorities, and is strongly supported by partnership arrangements. Later developments will depend on outcomes of these activities and will proceed in the directions identified by client countries.

To achieve even these modest initial measures, the Bank will need to undertake a number of specific internal measures: (a) updating its OP framework; (b) strengthening analysis to ensure that the Bank's ability to adequately consider forest impacts and issues in Country Assistance Strategies (CASs) and poverty reduction strategy papers (PRSPs) is enhanced; and (c) increasing enhanced loans and credits in the sector, based on a country's demand and readiness to invest in sustainable forest conservation and management. The objective will not be to necessarily have stand-alone forest operations, but to integrate forest issues in the broad agenda of country dialogue, poverty reduction strategies, and economic support.

Implementation of the proposed approach has significant implications for the Bank. Steps that will need to be taken include

- Modifying the 1991 Forest Policy to focus on protecting environmentally and culturally critical forest areas and permitting more proactive support of SFM (the new OP was approved by the World Bank's Board of Directors on October 31, 2002);
- Developing demand through knowledge sharing and country-driven ESW;
- Working with borrower governments to catalyze engagement and economically viable, environmentally sustainable, and socially responsible investments by the Bank, other donors, and the private sector;
- Building and strengthening partnerships;
- Applying selectivity and sequencing to the Bank's engagement, based on the principles outlined in chapter 4 (see especially table 4.1).

Modifying the Forest Policy

One of the most controversial issues relating to the Bank's operations in the forest sector is the impact of the Bank's current 1993 Operational Policy on Forestry (OP 4.36). In reviewing this policy, the OED, most participants in the regional consultations, the World Bank sector boards, and the Environmentally and Socially Sustainable Development (ESSD) Council recommended that the current Forest Operational Policy be modified.

By changing the OP, the Bank recognized the reality that few countries in either the developed or the developing world have been willing or able to devote more than 10 to 20 percent of their forest areas to strict Protected Areas and that, irrespective of any actions by the World Bank, most accessible and commercially valuable forests around the world will continue to be used for timber production sooner, in all likelihood, rather than later. The Bank also recognized that, frequently, the incentives for nonsustainable logging are strong for both governments and the private sector. Very often, the real choice available is not between doing logging and doing something less invasive and damaging to the forest, especially when these alternatives have not been developed to large-scale market viability. Rather, the choice is between doing logging reasonably well or doing it destructively so that conversion of the logged-over site to other nonforest uses becomes almost inevitable. The pace and proximity of agricultural and other forms of development can also influence this decision, but in any event it is essential that the questions of what forests are used for, and by whom, are considered rationally when making it.

The Bank believes it has significant comparative advantage in proactively assisting its client countries to improve the quality and sustainability of forest operations; address illegal logging; direct more of the benefits from forestry toward local communities; and ensure that logging operations are carried out in a manner that minimizes their impact on biodiversity, water catchment, and other forest environmental functions. Under no circumstances will the Bank Group invest in nonsustainable commercial logging or logging in environmentally or culturally critical forest areas, and this is reflected in the new OP.

Revisions to the Bank's (OP 4.36), discussed in chapter 2, will permit the Bank to become more proactive in both identifying and protecting critical forest conservation areas and in supporting improved forest management in production forest outside these areas. These revisions will retain many of the strengths of the current OP 4.36 and expand its effectiveness by using new initiatives in third-party certification and monitoring that have been developed since the previous OP was adopted in 1993. This new approach requires all forest harvesting and management operations financed by the World Bank to be monitored through independent assessment and certification. These third-party monitoring and evaluation systems have been developed through various initiatives and organizations involving civil society and the private sector. The risks associated with this approach are discussed under the general section on risks toward the end of this summary.

In addition, all Bank-supported forest harvesting and management operations are required to conform to existing Bank policies, including, especially, OP 4.01 (Environmental Assessment), OD 4.20 (Indigenous People) to be superseded by OP 4.10 (Indigenous People), OP 4.04 (Natural Habitats), OP 4.11 (Cultural Property), and OP 4.12 (Involuntary Resettlement). The implementation of this full complement of operational policies will provide the bases for ensuring that forest operations meet the high standards demanded by many stakeholders in the sector. Monitoring by government bodies, the Bank's own supervision, and independent third parties will provide the necessary assurances that forest operations are benefiting forest-dependent peoples while contributing to economic growth and poverty reduction. Although shortfalls in implementation have occurred in the past, the Bank has strengthened its capacity to meet the high standards that it has imposed on its operations. These standards continue to set the benchmark to which other organizations in development strive to achieve.

Developing Demand: The Global Commons and Economic and Sector Work

The Bank has recognized forests as a global environmental commons and has elevated conserving and managing forests well to be one of its global priorities. To meet this priority, the Bank is committed to improving both the level and effectiveness of its engagement in forests and increasing and strengthening diagnostic and analytical work within the Bank on forest sector issues. Unfortunately, forest ESW has declined precipitously in the Bank in recent years. The new strategy proposes increasing ESW as a primary instrument for enabling the Bank to reengage more effectively with borrowers and other partners in forest conservation and management. This will lead to increased interest in and commitment to sustainable conservation, management, and development of the forest sector on the part of borrowers, other stakeholders, and elements of the Bank itself. Incorporating timely and accurate analysis of the importance of forests into economic and social outcomes will be fundamental to the Bank's achieving its broader goals in poverty reduction and sustainable development. The Bank will need to do the following:

Provide the analytical support for effectively integrating forest conservation and management in CASs and into broader poverty reduction and economic development programs being prepared by Country and Sector Departments. Where there is appropriate support by borrower governments and Bank Country Departments, ESW will be undertaken both to identify the opportunities for sustainable forest development and conservation and to determine the potential impacts on forests and forest-dependent peoples of nonforest sector activities contemplated by the Bank through investment, adjustment, and broad poverty reduction programs. For the most part, this ESW will be sponsored by Country Departments, supported by partnerships and, in some cases, cosponsored by sector units responsible for design of the broad programs. This approach is intended to be demand driven: it will rely on expression of needs as defined by borrowers and Country Departments. The larger question of how Bank management and the Board will ensure that these larger programmatic investments take specific account of natural resource issues and outcomes in their design is not addressed in any detail here; that will be dealt with in specific discussions at the Board on the possible conversion of OD 8.60 (Adjustment Lending) to an OP, or a revision of it in its present format.

Develop and maintain an enhanced forest sector lending portfolio in the Bank. An increased ESW program is indispensable for lending. It will need to be leveraged by strong interaction with borrower countries, donors, and other stakeholders with common interests in the forest sector. This emphasizes the country ownership focus of the Bank's new approach in this sector, without which it is unlikely that any significant change in either the forest sector portfolio or forest conservation and development outcomes will result from this work.

While the normal ESW budget of the Bank will finance most of the sector work, some of these incremental activities may be financed through corporate programs aimed at developing the organization's focus on global public goods. As discussed later, knowledge-building partnerships with other private and public organizations, including the Program on Forests (PROFOR), also will contribute to the enhanced knowledge of the forest sector and cross-sectoral impacts.

Catalyzing Engagement and Investment

For the Forest Strategy to be successfully implemented, sufficient financing will need to be made available to assist countries to realize their national forest objectives for both development and conservation. This financing will need to be available on blended terms, ranging from investment funds at market rates through to concessional terms, including grants. To achieve blended terms, the Bank will need to work together with other donors including the Global Environment Facility (GEF), bilateral development assistance agencies, NGOs, civil society, and the private sector (including "green" private investment funds) to create the right blend of lending and grant financing from multiple sources.

To catalyze this financing strategy, the Bank will have to be prepared to raise its own profile and participation and to make available its own loans and credits to finance forest operations in close cooperation with other institutions and entities. The proposal presented in this Forest Strategy is that the Bank will support additional lending for the sector by centrally defraying the costs of preparing forest projects and programs incurred by Regions and Country Departments. The longer-term objective will be to lower the presently high transaction costs of forest operations by bringing more countries to a situation where they can absorb more programmatic assistance in this sector. This approach will be implemented under country programs, based on the interest and commitment of borrowers, and Country Departments' willingness to commit to incremental lending from their own resources, after some initial encouragement through corporate funding to cover the higher initial costs of preparing investments in this sector. Attention will be given to the issue of how best to provide incentives for the Bank's Country Departments to initiate lending in difficult sectors in which the issues are important to development and the Bank's mission but the risks and costs tend to be higher than the norm.

As noted, incremental Bank lending needs to be supplemented by other sources of finances. The motivation and coordinating framework will be based on a shared agenda for forests, so that all groups are able to focus their inputs on the same basic set of objectives in the sector. In many countries, this framework will be provided by enhanced NFPs. It is important to note that this concept has been developed through an extensive intergovernmental process which, among other things, has stipulated that NFPs must be led by governments and developed with the participation of all stakeholders. These NFPs will address investments and policies within the forest sector but also account for activities outside the sector that may impact the success or failure of forest operations and policies. Over the next five years, funding for forest activities in countries willing to commit to NFPs or like processes will be expected to increase significantly. The disbursement of these funds will be handled in parallel but coordinated programs to support countries in realizing their forest-related objectives.

Partnerships

Partners, including a number of important donors, already have indicated their willingness to work together, both analytically and financially, to realize the objectives of better forest conservation and management at both the country and global levels. Working with its partners, the Bank will need to assist in developing a framework for policy and investments. In many countries, the Bank already is operating within CDFs to establish the overall development and policy agenda for a country. The proposed NFPs will rest within the umbrella of the CDF and form the bases for partnerships in the countries. This proposed Forest Strategy argues that additional Bank investment in fostering partnerships in the three following areas will be essential to underpin its own efforts to raise engagement in forests and contribute to building coordination and agreement on priorities for investments and policy reform:

Partnerships with other donors. An existing forests donor partnership, PROFOR, has moved to the Bank from its previous location in the United Nations Development Programme (UNDP), which because of rationalization of the UN system has decided to reduce its direct involvement in forests. PROFOR is an analytical, knowledge-building program, focused on forests and forest-dependent people, and supported by a group of bilateral donors. It has strong linkages to the NFP process, and therefore to country-managed development of priorities and strategies for forest-based development. The principal PROFOR donors have strongly supported moving PROFOR to the Bank. They see major advantages in linking the country-led analytical and knowledge-disseminating capacity PROFOR can provide to the Bank's comparative advantages: an ability to analyze broadly across sectors and at the macroeconomic level; to establish and maintain a dialogue with senior political and economic decisionmakers; to apply a range of lending and nonlending instruments to reform and development; and to convene public, private sector, and civil society stakeholders for these and other related purposes. The move to the Bank will also enable PROFOR to draw on the knowledge and expertise of other research and policy bodies, including those in the Consultative Group on International Agricultural Research (CGIAR) system, with which the Bank has close relationships.

Partnerships with NGOs. The World Bank/World Wide Fund for Nature (WWF) Alliance for Forest Conservation and Sustainable use is a well-developed partnership for sustainable forest conservation and use. However, the Alliance requires additional support to increase its ability to draw on the strengths of each institution and to mainstream dialogue on important issues, including independent certification and monitoring of forest operations. Another major NGO partnership is the Critical Ecosystem Partnership Fund (CEPF) with Conservation International, which supports the protection and management of particularly important areas of biodiversity. Each partnership is time bound, with the Bank's contribution highly leveraged in terms of resource flows.

Partnerships with the private sector. The CEOs Forum, chaired by the World Bank president, has initiated a dialogue among leading forest companies, NGOs, and the Bank. Through plenary meetings and working groups, it has debated major issues such as managing forests sustainably and controlling illegal forest operations. In Africa, it has led to a draft code of conduct that could have significant implications for how responsible logging companies conduct business in the region. The CEOs Forum needs to widen its contacts in the private sector, especially to intensify the dialogue with some large institutional investors that have expressed interest in working more closely with the Bank on investments in SFM in some key Bank client countries. The new Forest Strategy will build on the CEOs Forum to strengthen its engagement with responsible investors in the private sector. IFC and MIGA are crucial to this engagement.

Together, these partnerships will enable the Bank to apply more effectively its multisectoral approach, convening power, and access to economic decisionmakers as well as its instruments, such as adjustment lending, the guarantee instrument, CASs, and other approaches to the central objective of improving poverty reduction and economic sustainability outcomes from forests. At the same time, other partners will be able to use their comparative advantages and financing capabilities to participate in the development of a broad, cohesive program in forests in major forest countries. Importantly, the Bank will maintain a small, independent technical advisory group (TAG) to help monitor implementation of the strategy and provide advice on any potentially controversial proposals for Bank investment support.

The result of these partnerships is to bring to the forest sector both knowledge that is widely shared and accepted and financing that is blended from public and private sources. Combining knowledge and sufficient financing at a spread of terms and rates will provide powerful incentives to bring forests into the mainstream of sustainable development and to assist in maintaining their environmental services to countries and the globe. Exceeding a threshold of knowledge and financing will be the most critical step to reverse the negative outcomes of increasing poverty and environmental degradation that often accompany the exploitation of forest resources.

Selectivity and Sequencing

Raising the level of engagement of the Bank will need to proceed in a measured and prioritized manner, particularly given the shortcomings of the past identified by OED in its thorough review of the Bank's forest operations. Selectivity and focus across countries will be important, as will linking specific forest interventions with country assistance priorities and programs. Two steps are envisaged in this process.

An initial selection of countries on which to focus incremental ESW and potential follow-up investments will need to be made. Obviously, regional and country teams and their client governments will be the primary actors in this process. In this selection, between 6 and 8 countries are contemplated for major focus, with perhaps 10 more for some additional attention. If the Forest Strategy is approved and the additional resources are provided through various sources, preliminary analyses by Bank Regional staff indicate significant potential for expansion in lending for forest activities and policy interventions, well beyond levels projected under business-as-usual assumptions. Obviously, in practice, a sequenced approach to expansion will be required, so that progress on sector analysis, international cooperation, the establishment of rational sector objectives and priorities, and the receptivity of the borrower in each case could be assessed.

As each country will be at a different level of readiness—with public commitment, institutions, and policy at various stages of development—it is important to prioritize the type of engagement. In the main text (chapter 4, table 4.1), a hierarchy of engagement is suggested, including the types of activities and associated estimates of the transaction costs per dollar lent. As noted earlier, the high transaction costs for forest operations will require higher-than-normal budgets for the Bank to prepare investments in this sector. The analytical and consultative processes that will be financed under this strategy through increased ESW and the leveraging of ESW through donor partnerships will lead to better-designed projects with higher and more specific benefits to the community and country. The knowledge base will be built through sector work, experience from projects, and learning through monitoring. Once this knowledge base exists and sufficient agreement emerges from NFPs on the priorities and basic reforms needed, transaction costs can be reduced to levels more consistent with or even below other Bank operations.

EXPECTATIONS, RISKS, AND REALITIES

Large Tasks and Institutional Constraints

The Bank must acknowledge, clearly and unambiguously, that the tasks implicit in the three pillars of this strategy are extremely large, while the institutional and human resources that can be brought to bear on them within the Bank in the short term are constrained. Some additions to the present staffing base (see section on Expectations and Realities: Risks and Monitoring Progress, in chapter 4) may be made soon, but, even with these, the Bank will need to rely on its partnerships, and also on the success of a catalytic approach, to play a significant role in achieving the desired objectives.

In the particular case of the forest sector, this is almost an inevitable situation. The reality is that the flow of funds from private and other sources into forests in the developing world (and in many instances these funds will be focused on unsustainable and even quite destructive practices) will continue to dwarf whatever financing the donors and multilateral lending agencies as a group can bring into this sector. Moreover, the flow of funds into cross-sectoral or economywide activities that can have major impacts (not necessarily always understood or anticipated) from all sources will always remain considerably larger than financing going directly into the sector from donors and multilateral lenders. It is essential, therefore, that the Bank and its partners adopt approaches that will help generate demand for sustainable but economically viable activities from borrowers and other stakeholders and assist in providing the means to create an enabling environment for these forms of forest use. This, in addition to grant-based funding to support the necessary protection of forests that should not be subjected to any form of large-scale production, will offer the best prospects for success at a scale comparable to the size of the resource and its current problems and issues.

Outcomes

It is useful to consider the magnitude of impacts that could result from an effective international program to address the Rio Forest principles, and subsequent intergovernmental agreements on what is needed in forests, to gain some appreciation of the importance of the task. On the basis of some previous commitments by governments and other stakeholders made through these processes, and figures on the scale and magnitude of the forests' potential contribution to development and environmental protection discussed earlier in this paper, it is reasonable to suggest that a significant and well-directed effort among donors, NGOs, the private sector, and borrower country governments and stakeholders should aim to have some meaningful impact on the scale of forest-related problems over the next 5 to 10 years. The following individual goals are suggested as reasonable for the international community and other major stakeholders as a whole:

Poverty. Improve the livelihoods of 500 million people, most of whom are poor and dependent on forest and tree resources, primarily through community forest management and development of agroforestry.

This number, while large, is less than half the number of poor people estimated to be dependent to some significant extent on forests for their livelihood. This can be regarded as a subset of the MDG on poverty reduction.

Governance. Strengthen the institutional capacity to reduce the losses from illegal logging by US$5 billion per year and improve the management of forest concessions to increase government revenues by US$2.5 billion per year.

As noted in chapter 1, the estimated losses from failure to collect appropriate royalties and taxes from legal forest operations are costing governments about US$5 billion annually. Illegal operations probably cost them a further US$10 billion in lost revenues. Recovery of half the amounts currently lost through improvements in the capacity of governments and other stakeholders to collect revenues, raise more reliable sources of financing for forest operations, and control illegal operations would represent a significant achievement, if not eliminating the problem entirely. The Bank's experience in forest governance reform in Cameroon (see Essama Nssah and Gockowski 1999) is indicative of the significant level of improvement that can be obtained in this field through a cooperative effort between government and other stakeholders to apply the right measures.

Protection and conservation. Bring 50 million ha of forests into new Protected Areas and improve the management of 50 million ha of currently Protected Areas.

These protection outcomes derive principally from commitments governments themselves have made in recent years, finalized into "stretch" targets as outlined by World Bank President Wolfensohn at the United Nations General Assembly Special Session (UNGASS) for Review and Appraisal of Agenda 21 in 1997. A recent midterm review of progress with the World Bank/WWF Alliance for Forest Conservation and Sustainable Use indicates that there has in fact been significant progress with these targets, and it is likely they will be exceeded within the original 2005 time frame set by the president.

Sustainable forest management. Bring 200 million ha of global forests under SFM that is independently verified and certified. This target is also one of the "stretch" goals enunciated by the Bank president at the 1997 UNGASS meeting.

Progress with this objective is likely to be slower than anticipated in 1997. However, the area of forest under certification has expanded exponentially since then, from virtually nil to about 27 million ha worldwide. About 9 million ha of this total is in Bank borrower countries, of which some 3 million is in tropical forest. Given that there are on the order of 600 million to 800 million ha of natural forest that already are, or will soon be, under some form of contractual or informal intent for production activities in the Bank's borrower countries, it would seem reasonable to aim at bringing 10 to 15 percent of this area under improved standards of forest management within a 10-year period. Fulfilling this aim, along with continued rapid progress in developed country certification, would achieve the target by then.

Risks

The final section of chapter 4 discusses the main risks of adoption of this strategy for the Bank and the mitigating measures that have been incorporated to deal with these risks. In brief, these are as follows:

Modifying the Forest Policy. Revising the OP, especially to alter the broad-based ban on Bank support for logging in primary tropical moist forest to more carefully targeted approaches, has been interpreted in some quarters as "opening the floodgates" to Bank support for unsustainable and destructive practices.

This is untrue, and the realities of the Bank's intentions in this regard are discussed at length in this paper and will also be reproduced in a Questions and Answers paper that accompanied the proposed revisions of the OP when it was publicly posted. This issue is discussed extensively in chapter 2 of this paper, which makes it clear that the intention of modifying the policy in this area was to ensure that the Bank becomes an effective player in the management of forests in an appropriate manner and to utilize various new approaches that will allow this to happen. There has been no financing of unsustainable logging by the Bank in Regions where the ban in the previous policy did not apply (since it applied only in primary tropical moist forest). There is in fact no indication of borrower demand for Bank financing of such activities. Bank support for commercial harvesting activities, in any event, is restricted in the new policy to situations where there will be independent validation of agreed standards in forest operations.

Credibility of program. There is some risk that some observers will conclude that expectations created by the large tasks listed under the three pillars of strategy, and the international community's overall Forest Policy objectives (discussed immediately above), cannot be achieved through the modest program of reengagement that is proposed in this strategy, and that there is therefore a major credibility gap. A related risk is that borrowers may not demand the assistance that could be made available in this sector.

As noted above, the reality is that the flow of funds from donors and multilateral lenders into forests, for management and protection purposes, will continue to be dwarfed by investments in activities that may have damaging impacts on forests. The Bank will need to continue to communicate the view that the only available approach is a catalytic one, aimed at building an enabling environment for sustainable use and effective protection of forests. This will aim at providing an appropriate enabling environment for financial support for sustainable forest conservation and management from other sources. So far as country demand is concerned, it is emphasized throughout this strategy that the ESW-led approach, and its particular focus on developing consensus and analytical support within countries for necessary reforms, is the best means to develop country ownership for this approach.

This risk will be further minimized by (a) ensuring that analytical work and other ESW are concentrated on building momentum and consensus for reform, not on analysis for its own sake, with flexibility so that resources can quickly be rechanneled to more promising prospects if the dynamic for change is not building; and (b) working closely with partners early in the analytical process so that understanding and confidence through which investments will be successful could be built jointly, through strong interaction with borrowers and other stakeholders.

Insufficient incentives. There is a risk that the measures proposed for reengagement in this strategy may not contain sufficient incentives to overcome internal Bank reluctance to engage in this sector, as exemplified in the "chilling effect" identified by OED, and the high transaction costs that are documented in this paper for forest investments.

One of the reasons for pursuing a relatively modest and cautious reengagement is to allow staff and managers to reduce their exposure to risk while re-engaging, first through the ESW approach, and then to investments if and when the necessary commitment from stakeholders is built up. The revised OP will provide a new and proactive atmosphere for forest engagement, and the emphasis on independent assessment and certification in this new approach to the management of forests also provides collateral monitoring of requirements of *other* safeguard policies, since these will enter the criteria and indicators for such assessment. Again, this will lower the risk of unforeseen problems with compliance with these policies. If investment in forests becomes more highly demanded as a result of this new approach, the Bank will need to consider corporate support for the higher preparation and supervision costs involved, to maintain the momentum of reengagement.

It is encouraging to note that at the time of writing, the Environmental and Rural Sector Boards in the Bank had initiated a program to hire more natural resource specialists in the Bank, and it is significant that the Regions have requested that an emphasis be given to hiring staff qualified to deal with forest issues and programs under this initiative.

Partnerships. The approach adopted in this strategy is heavily dependent on new partnerships, especially PROFOR, and these carry reputational and institutional risks for the Bank.

As detailed in chapter 4, a number of specific measures will be negotiated with the PROFOR donors, prior to establishment of the five-year program in the Bank, to clarify accountabilities, responsibilities, and continuity of support for this initiative.

The risks of a program aimed at these outcomes are considerable, but manageable. It should also be pointed out that the risks to the Bank of *not* adopting a more proactive stance toward forests are also considerable: There has been a strong expectation among most stakeholders that the Bank will reengage in the sector and will revise its policy to allow this to be done effectively. There is increasing interest and attention being paid to the general issue of the impacts of nonforest investments on forests, and at the very least the Bank must have an approach that will allow some of the relevant issues, and offsetting measures, to be analyzed and made ready for timely incorporation into the design of these large investments: the reputational consequences of inadvertent damage will be high.

Therefore, within the Bank there is now a major requirement for increased attention to quality and compliance, not just of forest operations but of *all* investment programs that could indirectly impact forest resources. The costs are equally manageable, even under tight budget constraints: the increased budgets proposed for the program in forests proposed above will amount to less than 0.1 percent of the Bank's overall budget. Critical, however, will be putting in place the necessary incentives to address difficult issues and manage risks and costs. Meeting this challenge is possible but requires some bold approaches to how the Bank does business. It also requires the development of support from governments, the private sector, and civil society working in partnership to achieve these results. A program that addresses economic priorities, addresses the well-being of the poor, seeks the voices of all stakeholders, and is open and transparent in its operation will be essential.

Monitoring

Monitoring of implementation will be an essential ingredient of success. The monitoring of the Bank's program will focus on four indicators:

Developing demand. The effectiveness of incremental analytical and consultative activities will be monitored in the focus countries. Criteria will include the alignment of NFPs and like processes with the major pillars of engagement put forward in this strategy, the establishment of working partnerships at the country level with other donors and major local stakeholders, and the degree of ownership and consensus on major elements of reform in the sector, including the development of interest and participation by major agencies of government in the agenda. Monitoring will be through analysis of programs and focus groups.

Building sector engagement. The test of this component will come later in the program, in the form of initiating the preparation of investments that have resulted from the ESW activities supported. In addition to normal Bank quality-at-entry and performance indicators, an important criterion will be the assessment of the degree of parallel financing or cofinancing by donor partners and the private sector that emerges as a result of collaborative activities. Establishing a baseline at the beginning of the program will be an important measure to allow assessment of this factor. Working with the Collaborative Partnership on Forests (CPF) will provide a base for monitoring financial flows. The amount and direction of financing flows to the forest sector in countries engaged in forest programs that are consistent with the three pillars of this strategy will be an indicator of implementation.

Impact of broader involvement on forest outcomes. It also will be necessary to evaluate the impact of the knowledge and analysis generated through this strategy on Bank activities originating outside the forests sector: the identification and incorporation of critical forest issues into cross-sectoral, adjustment, and broad programmatic lending and the inclusion of forest topics in broader ESW and CAS development work programs. Monitoring will include indicators of inclusion of forests in CASs and PRSPs.

Selectivity and sequencing. Effective disbursement and implementation of the incremental activities proposed, in the time frame outlined, will be a major factor here. The balance of activity types and progress toward the lowering of overall costs of doing business in the sector, through a movement of country situations being addressed up through the hierarchy of engagement as illustrated in table 4.1, will be a major factor for evaluation. Using table 4.1 as a model, indicators of types of engagement and costs will be recorded.

Although these indicators are important to track progress, effective monitoring will involve external stakeholder and technical advisers working with both Bank senior management and Committee on Development Effectiveness (CODE) engagement. Without some form of accountability that penetrates to the country level and teams, monitoring will be an empty exercise. Management will bear the responsibility to see that the overall objectives of the Forest Strategy are being realized.

External Advisory Group

External assessment will be built into the evaluation process, continuing the approach used during the preparation of the forest sector strategy paper (FSSP). Stakeholders will be consulted at every stage. Specifically, an ad hoc external advisory group (EAG) will be formed, utilizing the former TAG formed to assist the Bank with development of the new strategy and policy as a roster for selection on an as-needs basis. Small teams (8 to 10 members) would be formed for specific tasks, comprising individuals from major stakeholder groups (client governments, civil society groups, academic and private sector interests, and major donor partners) represented on the former TAG. The EAG process would be formulated in fiscal year (FY) 2003, in time to allow selected group(s) to assess design, relevance, and progress of the activities for implementation outlined below, with a specific objective of advising on implementation matters related to the Bank's OP and the overall aims of the strategy as set forward in the FSPP. In addition, the EAG will regularly advise the president of the World Bank regarding its view of the status of the forest program.

CHAPTER ONE

Challenges and Realities in Forests

IMPORTANCE OF FORESTS

Important progress has been made in improving overall living standards in a significant number of developing countries. Over the past 40 years, child mortality rates in developing countries have dropped by more than half, and malnutrition rates have declined by almost one-third. Despite these positive trends, poverty persists in many countries: 1.2 billion people live on less than US$1 a day; 2.8 billion people live on less than US$2 a day. Mounting evidence demonstrates that poverty—especially in rural areas—can be reduced only by sustainably managing natural resources both for the income they generate and for the environmental services they provide. The forests of the world are one of the most important of these natural resources.

The link between forest outcomes and poverty may not always be direct, or even evident in some cases. The economic benefits of forests are frequently undervalued, and they also often bypass the poor, because of existing tenure and participation conditions. Undoubtedly, in some cases, the poor would benefit from removal of forests and conversion to land uses to which they have more access, and in some of these cases, this conversion would be justified from a broader economic and environmental perspective as well. However, decisions on these matters must recognize the longer-term implications of change to natural systems, especially for the poor, who typically depend more upon their maintenance than do many others. There is considerable potential for the poor to benefit much more from all types of forest use, but this will occur in fact only if the right policies, institutions, and implementation capacity to include them are in place.

In general, forests provide support for nearly half of the 2.8 billion people living on US$2 or less a day. Thus, forests can—and must—take a far larger role in meeting the United Nations 2000 Millennium Development Goal of halving extreme poverty by 2015. A World Bank Group Forest Strategy must give priority to poverty reduction, while also focusing on lowering the wider risks of environmental degradation, biodiversity loss, and global warming. To meet these priorities, the Bank's strategy must focus on people as well as trees. It must participate in bringing local, national, and international stakeholders together to make decisions as to how forests are to be protected and used. The strategy also must recognize that forests are always a part of larger economic, environmental, and governance systems that must work together for the goals of poverty reduction, sustainable economic development, and environmental protection to be met.

Global agreements and initiatives require the World Bank to bring forest issues more deeply into the poverty reduction and sustainable development agendas. The 1992 Rio Earth Summit process has closely linked environmental concerns with sustainable development. New financing instruments are emerging that can considerably change the landscape for financing of goods such as biodiversity and

BOX 1.1

Why Forests Are Important

Forests cover 33 million square kilometers—26 percent of the Earth's land surface. They fulfill major economic functions, help maintain the fertility of agricultural land, protect water sources, and reduce the risks of natural disasters such as landslides and flooding. The world's forests are home to at least 80 percent of remaining terrestrial biodiversity and are a major carbon sink that mitigates climate change.

More than 1.6 billion people depend to varying degrees on forests for their livelihoods. About 60 million indigenous people are almost wholly dependent on forests. Some 350 million people who live within or adjacent to dense forests depend on them to a high degree for subsistence and income. In developing countries, about 1.2 billion people rely on agroforestry farming systems that help to sustain agricultural productivity and generate income. Worldwide, forest industries provide employment for 60 million people. Some 1 billion people worldwide depend on drugs derived from forest plants for their medicinal needs.

In the 1990s, forests were lost at the rate of 15 million ha to 17 million ha per year, and in some countries up to 2 to 3 percent of forest cover was lost per year (United Nations Food and Agriculture Organization [FAO] 2000). In some countries in the Asia-Pacific region, forest destruction is responsible for 2 to 5 percent per decade of global biodiversity losses, with inestimable losses to ecosystem stability and human well-being. Deforestation also accounts for up to 20 percent of the global greenhouse emissions that contribute to global warming. Mismanagement of woodlands in humid and subhumid tropical countries significantly contributes to soil losses equivalent to 10 percent of agricultural gross domestic product (GDP) per year.

Forests are consistently and seriously undervalued in both economic and social terms. For example, in Indonesia, official data show that forests contribute 1 to 2 percent of GDP, whereas the Bank estimates that the potential value of forests to that economy is closer to 15 to 20 percent of GDP. Nationally and regionally, forests provide important watershed, soil management, pollination, and pest management functions that usually are not captured by markets, in addition to timber and nontimber forest products. For many peoples, forests also are an important part of their cultural and religious heritage and practice.

While extremely difficult to quantify, the economic value of the ecosystem services of the world's forests is vast. A 1997 study in the journal *Nature* estimated the global value of the goods and services that forest ecosystems provide—from timber to climate regulation to water supply to recreation—at some US$4.7 trillion a year, more than one-quarter of that year's world gross national product (GNP) of US$18 trillion (Costanza and others 1997). The authors estimated that at least 70 percent of these values are generated in developing countries. Subsequent comments published in *Nature* and elsewhere questioned the basis used to estimate these aggregate figures by Costanza and others. Among the criticisms were, first, nonmarketed values from forests tend to be site specific; thus, aggregation, as used in the article, is problematic. Second, the basic usefulness of valuing the stock of forests is questionable since what matters ultimately are the changes in human well-being that result from incremental changes in the use of specific forest areas.

However, there is general agreement that the value of forest ecosystem services that are outside formal markets is significant. It also is generally agreed that relatively little account is taken of these values in much of the land-use decisionmaking that drives forest change. The challenge for policymakers is to bring these values into markets, cross-sectoral decisions, and macroeconomic policymaking and into the development of the economy in general.

Sources: Costanza and others 1997; FAO 2000.

carbon, both of which are integral to forests and their management. The World Bank is prepared and obligated to assist its borrowers to meet their commitments under international conventions and to take advantage of new financing opportunities to manage their forests.

International Context: Global Conventions and Agreements

The legal and international framework that governs forest issues has advanced and broadened since the last Bank Forest Strategy was issued in 1991, bringing forth additional concerns and providing new opportunities. The World Bank Group's new Forest Strategy must be consistent with and support this emerging international framework.

Rio Earth Summit. The 1992 Declaration of Forest Principles at the Rio Earth Summit affirmed that states have sovereign rights over their natural resources, but also recognized that forests are a global public good that provides ecosystem services of global value and significance, such as biodiversity preservation, carbon sequestration, and nutrient and hydrological cycling. Throughout the 1990s, forests became a topic of increasing concern, contention, and cooperation in diplomatic circles and in the global marketplace. In October 2000, through the Economic and Social Council of the United Nations (ECOSOC), the international community created the United Nations Forum on Forests (UNFF), a new international body that will provide a platform for high-level policy discussions and cooperation. The Bank is obligated to assist its clients to meet the commitments and international conventions arising from the Rio Earth Summit.

Kyoto Protocol. The United Nations Framework Convention on Climate Change's (UNFCCC) 1997 Kyoto Protocol created three "Flexible Mechanisms" to mitigate climate change. Two of these relate to the Bank's client countries. The first, "Joint Implementation," allows parties within the "Appendix 1" countries (developed countries and countries in economic transition) to transfer or acquire emission reduction units from any other party. This mechanism could play an important role to support SFM in countries in economic transition. The second mechanism, the "Clean Development Mechanism (CDM)," regulates greenhouse gas emission trading between industrial countries and developing countries. The role of forests in CDM has yet to be fully defined; however, the protocol holds out the possibility of integrating forest management and conservation with a particular emphasis on reforestation and afforestation activities. This integration could mobilize substantial resource flows to developing countries. The Bank is not directly engaged in the political debate that underlies the Protocol but has taken up the role of testing the market for some of the flexible mechanisms to ensure that this debate is informed by practical experience. In addition, a Bank representative has chaired the scientific panel examining the evidence on global warming. Because of both the potential resource flows from markets in carbon and the need for its clients to adapt to possible climate change, the Bank has a strong role to play regarding climate change, especially in the forest sector.

GLOBAL CHALLENGES AND OPPORTUNITIES

Many studies have drawn attention to the formidable global challenges presented by forests. These studies underscore the point that the forest sector represents one of the most challenging areas in the development of community and global public policy.

Despite significant resource flows, international concern, and political pressure, the potential of forests to reduce poverty, realize economic growth, and be valued for their contributions to the local and global environment has not been fully realized. A combination of market and institutional failures has led to forests failing to contribute as significantly to address these issues as would be possible under good economic and technical management. Instead, the forest sector often demonstrates the failure of markets and governance to capture its full value.

Failure of governance. Forests often have been disregarded in economic policy and have been plundered for short-term gain and, frequently, then removed and replaced by less valuable and less sustainable activities. Such loss with degradation often has been at the expense of national economies and rural people who depend on forest resources for their livelihoods. Despite their desperate need for fiscal resources for development, governments have failed to capitalize on the value of their forest resources. World Bank estimates place the annual revenue loss to governments from failure to collect taxes from forest concessions at more than US$5 billion. In addition, the annual market value of losses from illegal cutting of forests is placed at over US$10 billion.

Need to integrate forests in the fight against rural poverty. To realize the international commitment to halve the number of people living in absolute poverty by 2015, increased prosperity must be brought to rural areas, in which the majority of the poor live. In many developing countries, rural poverty cannot be reduced unless forest resources can be sustainably developed and better used. Broadly defined, forest resources include dense forests, open woodlands, agroforestry, smallholder woodlots, and commercial-scale plantations. However, rural strategies often have neglected forests, because the latter are mistakenly viewed as outside the mainstream of agricultural development.

Need to deal with conflict in resolving forest issues. Forest Policy has become one of the most controversial and heated issues in development. To use forests for poverty reduction requires a strong institutional framework—and an effective legal and regulatory environment—in which the rights of specific groups among the poor are recognized and protected, while opportunities to develop sustainable forest businesses are provided to these people and other groups. Such a framework often has been lacking. Dealing with these issues puts a premium on participation, conflict resolution, and an institutional structure attuned not only to the technical and economic issues in forestry but also the inevitable conflicts.

Failure of markets to capture environmental services of forests. Forests deliver some products—primarily lumber and fuelwood—through markets. However, many of forests' other contributions to the environment, biodiversity, and the stability of the global climate are not recognized in terms of financial values and, therefore, go unrecognized in markets. Although rarely estimated, the indirect consequences for a nation's economy and environment of this undervaluation are likely to be even more devastating as local people lose their sources of fuelwood and fodder and the protection that intact forests offer their water and soils.

Need to account for global values from forests. Globally, forests are one of the world's most manageable sinks for carbon and the home of most of the planet's terrestrial biodiversity. Because these values contribute to the world's well-being, they are not only local values. Climate change will have widespread consequences for both the poor and the rich. A world without significant diversity of life has negative economic consequences and will impoverish the lives of all people.

To maximize its potential to address these challenges, the Bank will have to focus its efforts—both those specific to forests and those it undertakes more broadly that can impact forests and forest peoples—on the roles of forests. Bank activities must facilitate forests' role in reducing poverty, must integrate forests in sustainable economic growth, and must protect the local and global environmental and cultural values that forest ecosystems provide. In making these efforts, the Bank will need to function at both the country and global levels. Most importantly, its contribution will need to be built on partnerships, consultation, and participation. Unfortunately, despite its visible economic and political presence in many key forest countries, to date, the Bank has not always been able to rise to these challenges.

WORLD BANK FOREST PERFORMANCE, STRATEGIES, AND POLICIES

These values of forests and the institutional and market failures in managing the sector have not gone unrecognized in World Bank strategies and policies, although the emphasis and remedies have evolved over the years, guided by major policy papers published in 1978 and 1991 (box 1.2).

OED Review of the 1991 Strategy and Policy

In a 1999–2000 review of the Bank's 1991 policy, the OED lauded the Bank's ambitious goals and the strong signal the policy had sent about the Bank's new emphasis on conservation. However, OED found that implementation of the strategy had fallen far short of these goals and that the logging ban and controversy about the policy had had "a chilling effect" on new forest initiatives. OED asserted that the Bank's direct forest lending had stagnated, "hobbled" by low borrower demand, high transaction costs, and fear of public controversy. OED termed the effectiveness of the strategy "modest" and called for a new policy that would "make the Bank Forest Strategy more relevant and strengthen the Bank's ability to achieve its strategic objectives in the forest sector." OED called for the Bank to integrate forests more closely in its overall mission of reducing poverty and to bring forest strategies into rural development programs.

OED also recommended that the Bank take a much more active role in forest-related activities, extending its reach by mobilizing additional concessional resources for the sector and forming partnerships with national and global stakeholders. The department also contended that the Bank should broaden its role away from concentrating on moist tropical forests to encompass all types of natural and planted forests and that conservation efforts should

BOX 1.2

World Bank and Forests

The World Bank financed its first forest operations—in then Yugoslavia and Finland—in 1949. The early projects financed the purchase of timber equipment. The Bank's approach evolved from a focus on industrial operations to social forestry and agroforestry, then to an emphasis on conservation. Prior to the publication of its 1978 Forestry Sector Policy (World Bank 1978), the World Bank Group had supported forestry primarily by investing in log extraction operations, pulp and paper mills, and technical assistance for species trials and by strengthening forestry institutions. Forest-related lending was ad hoc and lacked overall appraisal of the potential of forests to contribute to economic development or environmental protection. In addition to the role that forest industries play in creating employment and generating incomes, the 1978 policy paper reflected growing awareness of the ecological functions of forests and their contributions to agricultural productivity and enhanced rural incomes.

During the 1980s, rising international concerns about the escalating rate of tropical deforestation and its implications for biodiversity loss, global warming, and environmental degradation triggered a revision of the Bank's policy. The ensuing Forest Policy paper, "The Forest Sector" (World Bank 1991e), recognized poverty reduction, policy reforms aimed at containing deforestation, and resource expansion as themes. At the same time, the policy paper strongly emphasized preserving intact forest areas and included a Bank commitment not to finance commercial logging in primary moist tropical forests under any circumstances. The broad goals of the 1991 strategy were to prevent or significantly reduce deforestation and to stimulate plantations and creation of additional forest resources. The strategy argued that the existence of major externalities arising from forest use and a range of perverse policies and incentives were leading to exploitative and unsustainable use of natural forests by a range of stakeholders—including the poor, who were seen as a major cause of forest destruction.

Source: World Bank data.

be streamlined and linked to national development goals. OED called for the Bank to mainstream its forest activities by giving them more weight in macroeconomic analysis and paying greater attention to forest issues in its other sector operations. The department advised that the Bank should provide its staff adequate resources and incentives "to address the risky and controversial issues of the forest sector." It also pointed out the impact that national governance issues have on forests—particularly regarding illegal logging—and recommended that the Bank mobilize national stakeholders to improve and monitor governance. Finally, OED called for the Bank to continue its involvement at the global level through participation in the United Nations' efforts to unite governments to achieve better forest management.

OED's analysis included a survey of Bank staff involved in forest issues. The survey found that staff tended to agree with the overall aims of the 1991 policy but saw problems in its specifics and did not believe that the Bank had succeeded in its major aim of reducing deforestation (box 1.3).

Appendix 6 (on CD) adds detail to OED's data and summarizes the history of the Bank's forest portfolio from 1992—when the new strategy and policy began taking effect—until 1999. The "chilling" effect identified by OED is evident: over the 1992–99 period, direct Bank investments in forest management in the tropics either declined or remained static at fairly low levels. Appendix 1 summarizes the responses to OED's recommendations that are embodied in this revised strategy.

Additional Indicators of Performance

Appendix 9 shows strong regional differences in investment in forests and in specific responses to sector issues. The data also indicate that the Bank's forest portfolio in general is in decline. Disbursement trends show a decrease since 1995, and the Bank's forward pipeline implies a continuing decline in forest sector projects, unless the priority given to forests changes. Very few forest projects are in preparation for Africa or Latin America, and problems exist elsewhere as well. ESW in the forest sector, a

BOX 1.3

Staff Survey: Effectiveness of 1991 World Bank Forest Policy and Strategy

The Bank's OED survey of staff concerning the 1991 Forest Policy found that:

- The ban on Bank financing of commercial logging in primary tropical moist forests had no impact in these areas. A more flexible policy that encouraged the Bank to enter this realm (on the correct side) would have been more effective.
- Forest sector issues were not well integrated in the Bank's broader mission of poverty reduction and economically and socially sustainable development. Performance in the sector would be improved by promoting more focus on natural resource protection, institutional reforms, and multisectoral approaches to forest development and plantations.
- Country managers saw forest sector involvement as high cost, low return, and risky. Apart from the institutional risk factor, the staff believed that there were internal Bank reasons for this managerial attitude. These reasons included inadequate resources ESW, project preparation, and supervision—all of whose costs have been relatively high. This past experience reflects the reality that, in many situations, politically difficult land use and social and environmental issues have had to be addressed in the course of project design.
- Staff also identified exogenous reasons for managerial reluctance to become engaged in forests. These reasons included the high levels of corruption often present in the sector and inadequate appreciation of the sector and its issues by policymakers in client countries. The latter was exacerbated by the low status and influence of forest and environment ministries in most cases.
- The Bank could and should become a global leader in forest-related matters, such as climate change, carbon and the Clean Development Mechanism, biodiversity conservation, and natural resources management. A clearer strategy and incentives, with the necessary resources, would allow better use to be made of GEF resources.

Source: World Bank 2000g.

major prerequisite in building a portfolio pipeline, fell from US$3.2 million in 1992 to US$870,000 in 1999. This decline has been general across all regions.

Staff turnover and high costs also have been contributing factors to this decline. An average of 3.2 task managers has been engaged in each forest project, from preparation through close. It is well known and recognized in the Bank that a high turnover of task managers throughout a project is undesirable for continuity and good supervision. The average cost of preparing and supervising a forest loan is approximately 1.6 percent of the value of the loan; the Bank project loan average is on the order of 1.1 percent. This added budgetary cost has been a disincentive to Bank managers to engage in the sector. Some of this difference reflects the fact that forest projects tend to be smaller than average, and some preparation costs are largely fixed regardless of project size. Nevertheless, the bulk of the difference is due to the complexity and high transaction costs of dealing with consultations and safeguards in this sector. Clearly, there is a need to deal with these costs through seeking greater efficiencies, selectivity, and reduction in the average costs of forest lending operations.

New Directions in Bank Priorities

As elaborated in chapter 4, the Bank has finalized the selection and definition of its global priorities and is in the process of better aligning global corporate priorities with country program goals. Forests are included in the top five issues singled out as integral components of the Bank's new global priorities: the environmental commons.

The principles underlying the strategic approach to these priorities include effectively mapping them into country programs, applying realistic stretch targets for individual priorities, and developing the action programs and cost estimates to do so. These principles are adopted for forests in this strategy. In the strategic approach to the global priorities, the need to centrally budget and finance these priorities is recognized as an incentive for implementation at the country level. Chapter 4 gives specific attention to this aspect of implementation of the new Forest Strategy. Sector strategies are to focus on the priorities and the needed programs and targets to achieve them. CASs are to be aligned with the priorities, and sector boards are to monitor implementation. To ensure alignment with its global priorities, the Bank will need to apply a multisectoral approach to forests and integrate forest priorities and issues in its broader CDF and its major poverty reduction initiatives, such as PRSPs.

LINKS TO OTHER BANK STRATEGIES AND POLICIES

Bank Safeguard and Operational Policies

For the Bank to implement a genuinely multisectoral approach to forests, it would have to address the policy, institutional, and structural issues in broader nonforest sector programs that have particular influence on forests and integrate forest outcomes in these programs. The current review of OP 8.60, which governs the implementation of structural adjustment programs, will be of particular relevance in this regard.

Also relevant are the provisions of environmental assessment embodied in OP 4.01, which require that impacts of any proposed activity on the natural environment, human health and safety, and social aspects are taken into account, referencing Operational Directive (OD) 4.20 (Indigenous Peoples), and Operational Policies 4.11 (Physical Cultural Property, forthcoming), 4.12 (Involuntary Resettlement), and 4.04 (Natural Habitats). OP 4.04 in particular requires that the Bank not support projects that, in its opinion, involve the significant conversion or degradation of critical natural habitats. In cases in which conversion or degradation of noncritical natural habitats is necessary, the project must include mitigation measures acceptable to the Bank. These mitigation measures include minimizing habitat loss, for example, strategic habitat retention and postdevelopment restoration, and establishing and maintaining an ecologically similar Protected Area. The Bank accepts other forms of mitigation measures only when they are technically justified.

Integrating Forest Issues in Poverty, Environment, Rural Development, Gender, and Water Strategies

The Bank will need to continue and further exploit its comparative advantage to influence nonforest sector policies that impact forests. It also will need to proactively seek opportunities to integrate the main elements of this new Forest Strategy into the Bank's broader poverty reduction, rural development (including integrated natural resources management), environment, gender, and water strategies.

In December 1999, the Bank and the IMF endorsed a new framework for poverty reduction. This framework requires the preparation of PRSPs as a basis for providing funding from the Bank and IMF, as well as debt relief under the Heavily Indebted Poor Countries (HIPC) initiative. Recently, there was intensive dialogue within the Bank concerning how to ensure that these PRSPs incorporate and quantify environmental determinants of poverty dimensions. The results of this dialogue will be vital in forming a real link between forest outcomes and broad Bank activities aimed at poverty reduction. Chapters 2 and 3 and appendixes 2 and 3 provide some quantification of the economic importance of forests and their relationships to large numbers of rural poor. In the formulation of target country PRSPs, it will be important that the strategy and implementation plan for forests provide opportunities to factor forest information in the broader proposals for poverty reduction.

The rural development and natural resource framework that has been developed focuses on the two highest global good priorities for which the Bank's Vice Presidency for ESSD has responsibility: (1) environmental sustainability and protection of the environmental commons, which includes atmosphere (climate change and ozone depletion), biodiversity, forests, food and water security, and land management; and (2) long-term social sustainability, with emphasis on the empowerment of poor people, voice, engaging civil society in dialogue, community-based development, and managing pre- and postconflict situations. The approaches and issues raised in this proposed Forest Strategy are fully consistent with these broader priorities. This congruence will make integration of forest issues and concerns in larger rural development programs relatively straightforward, so long as the investment in linking the programs and strategies is made at the field level.

Similarly, this proposed strategy is consistent with the three basic objectives of the Bank's environment strategy: (1) improving people's quality of life by focusing on enhancing livelihoods and reducing environmental health risks and the vulnerability of the poor to natural disasters (such as deforestation-induced landslides and forest fires); (2) improving the quality of growth, with special emphasis on helping countries to develop a better policy, regulatory, and institutional framework for sustainable economic growth and on strengthening environmental safeguard systems; and (3) improving the quality of the global commons through approaches that include special financing mechanisms to compensate countries for the incremental costs they incur to protect the global commons.

As outlined in the Bank's "Integrating Gender into the World Bank's Work: A Strategy for Action" dated January 2002, one of the main components of the World Bank's strategy for changing gender patterns, which are potentially costly to growth, poverty reduction, and human well-being, is to work with client countries to prepare periodic, multisectoral country gender assessments. Country gender assessments are designed to analyze gender dimensions of development across sectors—including rural development and forest management—and to identify gender-responsive actions. As pointed out in numerous international reports, the sustainable use of forests requires the participation of all rural populations, including women. Although women's needs often differ from those of men, many programs tend to overlook women's specific needs regarding forestry—often because policymakers lack adequate data, information, and methodologies to address them. The lack of gender awareness constrains the sustainable use and management of forests and forest ecosystems in many places throughout the world. Under the new Forest Strategy, Bank staff will work to ensure that country gender assessments are appropriately designed to help identify gender dimensions of forestry and rural livelihoods in individual countries. Likewise, forestry-related analytical and advisory services will incorporate gender issues as part of its analysis. Together, these can be used to develop priority policy and operational interventions that respond to these assessments. At the project level, as identified in the Forest Strategy (e.g., Box A2.2), gender analysis will be an important tool to provide simple information on resource use, responsibility, perspectives, and needs and therefore serves a critical role in the quality of forest investment design. Gender analysis has already been incorporated as part of social assessments that are now routine during the preparation stage of forest projects.

The World Bank's Water Resources Sector Strategy focuses on the role that effective water resources development and management play in sustainable growth and poverty reduction. Improved water resource management (such as watershed projects in degraded environments) is a major focus of the emerging Strategy and is shared with the Forest Sector Strategy. There is a growing program in watershed management in the Bank portfolio at present, and this will necessitate closer programming and working relationships between forests and watershed technical and operational staff in the Bank. Strong links between the Forest Sector Strategy and the emerging Water Resources Sector Strategy will ensure a sharing of common issues.

DEVELOPING AND IMPLEMENTING THE NEW BANK APPROACH TO FORESTS

The Importance of Country Ownership

One of the major requirements of an effective reengagement by the Bank in forests in its borrower countries will be commitment by the borrower, and other major local stakeholders, to the objectives and means proposed by the Bank to do so. This might be seen by some as a major constraint on the speed at which a reengagement can proceed, from the low base that has been produced by the Bank's virtual absence from the sector in some countries. In fact, it is seen here as an opportunity in these countries to engage in a dialogue and an analytical process involving country stakeholders *before* a return to sectoral investments is implemented, through a concerted ESW activity. Thus, the implementation strategy for the Bank's reengagement in forests proposed in this strategy emphasizes enhanced ESW initially, which will allow this dialogue with governments and other stakeholders to commence or, where it is already in progress, to develop. Moreover, a focus on developing consensus at the country level through country-driven dialogue and analysis of major sector issues and priorities will also open the way for incorporation of sectoral concerns and measures that result in larger Bank investment activities, instead of these inputs being either introduced through a top-down-only approach or ignored completely.

A major reason for adopting the three pillars of strategy outlined earlier in this chapter is that they encompass the broader interests that borrower governments and other local stakeholders have in the management and use of their forests, by comparison with the more restricted focus that arose, in some cases, from the previous Bank 1991 strategy and 1993 policy in this sector. This is a deliberate choice, recognizing that the success of the new approach will be completely dependent on the development of demand for Bank involvement of various types by governments and other interest groups, and that this will be forthcoming only if these stakeholders see reality and relevance in the broad strategic approach the Bank adopts. The approach set out in this strategy relies on stakeholders' willingness to take a leadership role in the analytical and consultative work that underpins this approach, so that the results of this work genuinely develop country-level consensus and commitment to whatever lending or nonlending activities are eventually proposed for that country.

The partnerships and processes proposed in this strategy to implement the new approach have been chosen with a view to their compatibility with country ownership. In the case of the new PROFOR initiative, for example, it will be of central importance that activities it sponsors are determined by criteria relating to the Bank's own broad poverty and sustainable development agenda with its borrowers, since these are now broadly defined by countries themselves through the CAS process, and then increasingly refined in design by countries through PRSPs and other mechanisms. Equally, however, PROFOR will have close links to the NFP approach. This process was designed and approved through extensive intergovernmental dialogue on forests over recent years; it therefore already has strong political ownership by countries that participated in that process, and its design requires that individual country governments must take ownership of an NFP for it to progress.

Developing the Approach

Chapter 2 sets out the strategic pillars for the Bank's new Forest Strategy. In the process, it considers some potential difficulties or conflicts that need to be addressed when considering approaches. These challenges include

- Difficulties inherent in assisting one category of poor people living near forests without inadvertently worsening the situations of other groups;
- Controversy that surrounds the use of the concept of SFM as a basis to engage in operations in natural forests;
- Limitations on present approaches in the Bank to incorporate forest issues and activities in larger cross-sectoral and macroeconomic programs;
- Constraints on the Bank and other funding agencies to finance protection of forests, as compared with the magnitude of competition for forest areas;
- Options available to leverage the impact of these limited financial resources.

Finally, chapter 2 summarizes some projections of future activities to support forests that the Bank could consider, in the context of an expanded program of involvement in the sector.

Implementing the Strategy

Selectivity. Implicit in the discussion of the strategy is the principle of *selectivity* in Bank activities in forests. Country ownership will be a broad criterion in this determination. The specific activities discussed in chapter 2 indicate a refocusing of Bank activities in this portfolio on poverty reduction, with the resulting increased emphasis on forest resources and forest-dependent people. In addition, the need to integrate forest activities and issues in larger cross-sectoral and broad economic programs will progressively move the forest portfolio toward a more programmatic and less specific, project-oriented approach. How the Bank, borrower governments, and their partners should develop the forest portfolio, and in what way, also is discussed in chapters 2, 3, and 4. Chapter 4 makes an argument for selective engagement based on country commitment and readiness for support. This engagement could range from basic policy dialogue to large-scale programmatic lending.

Developing partnerships. As discussed in chapter 3, the World Bank has formed partnerships with other institutions—in part through an interagency task force on forests and now through a newly formed Collaborative Partnership on Forests in support of the new UNFF. The Bank also is building a specific partnership to work on forests with bilateral donors (PROFOR). In addition, the Bank has formed partnerships with NGOs—including the WWF, Conservation International, and other civil society groups—and is further exploring closer relationships with large, international, private sector investors interested in following a sustainable development approach to forests.

These partnerships are an important part of the Bank's new strategy. They will enable the Bank to build on its comparative advantages in addressing forest issues through a broad spectrum of cross-sectoral strategies and policies. The partnerships will permit the Bank to collaborate closely with national governments and other groups that will be more able to take up some of the specific technical, capacity, and knowledge development activities needed for an integrated forest intervention.

Coordinating across the Bank. Preparation of this revised Forest Strategy involved close collaboration between the ESSD Forest Team, which is responsible for its preparation, and Regional sector operational staff and managers (chapter 2). The IFC also was a participant in the strategy preparation. Chapter 3 argues that it is important that this level of cooperation and coordination be maintained in the future. The World Bank Group must follow the same strategic path to encourage sustainable development of the forest and forest industry sector. Its several institutional arms—the International Development Association (IDA), IFC, MIGA, International Bank for Reconstruction and Development (IBRD)/World Bank, and the World Bank Institute (WBI)—will work on the basis of a unified strategy to implement internationally supported and financed technical support and economic analysis.

Financing the new strategy and leveraging impact. As noted at the outset of this chapter, forests presently are regarded as an important element in one of the Bank's new global priorities: the environmental commons. Significant changes in approach to the sector are expected to result from this status. The financing strategy proposed in chapter 4 is based on:

- Encouraging the development of blended financing arrangements from multiple sources to reduce the overall financing cost to client countries of overall forest and environmental services. Developing blended financing would lead the Bank, donors, and the private sector to more closely coordinate their support. These funds would go to countries willing to implement national programs of SFM, Forest Policy reform, and improved governance and institutions. The IFC will continue to invest in private sector companies that are willing to operate in accordance with internationally accepted criteria for SFM and are committed to independent certification of their operations. For all projects involving the utilization of forest products or the protection or enhancement of environmental services, the Bank will require the design to include an evaluation of the prospects for the development of new markets and marketing arrangements for the full range of goods and environmental services that can be produced by well-managed forests
- Encouraging an enhanced commitment to focused ESW. The approach described in chapter 4 is to use some of a proposed increase in ESW to leverage larger increases in donor funds committed to policy development and consultative purposes into cooperative analytical work

CHAPTER TWO

The Forest Strategy: Proposed Bank Action

An effective Bank Forest Strategy must consider all activities, current and potential, that can influence forest and related social, economic, and environmental outcomes—rather than focusing only on forestry activities such as production or conservation. However, the Bank must be selective in its engagement, seek its comparative advantage, and build on the pillars that set the base for its activities.

PRINCIPLES OF ENGAGEMENT AND COMPARATIVE ADVANTAGE AND THE PILLARS OF THE STRATEGY

The Bank's impact on forests is determined not only by its direct investments in forests and forest-related activities but also by its investments in other activities that have indirect effects on forests and forest-dependent peoples. As a consequence, its nonlending activities, analysis, dialogue, and safeguard policies take on increased importance. In addition, to have impact on the ground, the Bank must follow the principles that underlie the CDF. These principles—partnerships, ownership, and a framework for engagement in the sector that defines the roles and requirements for progress—all will be fundamental to success.

Lessons for Engagement in Forests

Bank experience in the forest and other sectors suggests several principles that need to apply to its new strategy:

The Bank will help build and respect ownership. Governments and people must own the policies and interventions that will make their forests more viable. This ownership can be built only through knowledge, participation, and consultations. With its partners, the Bank must help bring knowledge to bear and be open in supporting the voices of the people.

The Bank will operate in partnership. The Bank has neither the financial resources nor the staff to meet the objectives of this strategy on its own. Specializing in its comparative advantage will enable the Bank to concentrate on its strengths without diffusing its resources, while drawing on and complementing the strengths of its partners.

The Bank's commitment to address forest issues will be long term and of a sufficiently large scale to enable change to materialize. The foundations of sustainable management of forest resources are good governance and institutions. Building these foundations will take time.

Comparative Advantage

The new Forest Strategy relies on the Bank Group to focus on its strengths. The Bank Group has comparative advantage in the forest sector in five areas:

1. *Cross-sectoral analysis, dialogue, and investments.* The Bank has the ability and means to operate across sectors with both policy dialogue and investment programs. Its economic and technical

work emphasizes linkages across sectors, with the macroeconomy, and with trade.

2. *Economic policy and country dialogue.* The Bank's leadership in country policy dialogues and assistance strategies gives it the capability to foster dialogue and influence policy. This dialogue has expanded to include support to good governance and combating corruption.
3. *Convening power.* The Bank is capable of bringing together stakeholders and donors to discuss important issues and to set objectives for assistance strategies. As the chair of the consultative groups for most countries, the Bank has the responsibility to work with governments and donors as they formulate their investment programs and policy agendas. This convening power allows the Bank to be catalytic in bringing forth new and experimental programs and agendas.
4. *Scaling up and mainstreaming projects and programs.* Through its ability to bring support through large loans and grants to projects and programs, the Bank has the means to scale up and mainstream successful projects and programs. In doing this, the Bank can encourage and support meaningful policy reforms and institutional changes. The Bank Group also can bring into play a multitude of financing and guarantee instruments to support the engagement of many different stakeholders, including civil society and private enterprise, and can address issues such as political and regulatory risk.
5. *Working with the private sector.* The focus will be on aligning public sector interventions and private sector objectives. Through its focus on the enabling environment for environmentally, socially, and economically responsible investments, catalytic investments by IFC, and the guarantees provided by MIGA, the Bank Group has an important role in stimulating appropriate private sector activity in the forest sector.

Three Pillars of the Forest Strategy

The Bank's Forest Strategy will be based on three equally important and interdependent pillars:

1. Harnessing the potential of forests to reduce poverty
2. Integrating forests in sustainable economic development
3. Protecting vital local and global forest environmental services and values

These three aspects must be addressed *together.* This necessity makes the Forest Strategy complex and multifaceted. It is not simply about growing or protecting trees but is a complex interaction of policy, institutions, and incentives. A narrow perspective on forestry—even sustainable forestry—is insufficient. To be effective, the strategy demands a multisectoral approach that addresses cross-sectoral issues and takes into account impacts on forests and forest-dependent peoples that originate from activities, policies, and practices outside the sector.

HARNESSING FORESTS' POTENTIAL TO REDUCE POVERTY

Economic growth alone cannot combat poverty effectively. More focused interventions are required that address opportunity, empowerment, and security and that acknowledge the potential conflicts inherent in addressing the different groupings of poor people who depend on forests in differing ways. A broader livelihood approach must be taken that places forests—productive capacity, institutional and legal structures, market access, and tenure—in the broader context of rural development. Priority areas for Bank action will include:

- Promoting policy, institutional, and legal frameworks that ensure that the rights of indigenous and other forest-dependent peoples and communities are protected
- Empowering women, the poor, and marginalized groups to take a more active role in formulating and implementing rural forest policies and programs
- Supporting the scaling up of collaborative forest management so that local communities can manage their own resources, rehabilitate and protect forests, market forest products, and benefit from security of tenure. The emphasis here will be on involving small farmers. A principal mechanism will be much closer integration of forest activities and investments in broader rural development programs
- Working with local groups, NGOs, the private sector, and other partners to integrate forest and agroforestry farming systems into rural development strategies

Appendix 2 (on CD) summarizes some past experiences of attempts that have been made by the Bank and other agencies to incorporate poverty-reduction concerns in forest strategies and national development plans. Appendix 2 highlights some of the more difficult issues and lessons learned from these experiences.

Ensuring participation of the poorest. Although most rural households are poor in absolute terms, there are considerable variations in levels of wealth among the rural poor, even in individual communities. This variation usually means that only the better endowed, or politically more powerful, are able to take advantage of the more rewarding forest production opportunities that are available. Moreover, all too often, their advancement is at the expense of the poorest, who may easily find themselves excluded from access to the resources on which they rely. As the intention of the new Bank Forest Strategy is to help the poorest as well as those better able to benefit from assistance, care will be taken to structure Bank support to reach all segments of rural society.

Protecting access by the poorest to nontraditional forest products. A primary focus of Bank support for many forest dwellers will be reducing poverty and ensuring continued access to subsistence supplies. Specific measures through which the Bank will seek to implement this strategy include:

- Implementing the policies spelled out in OP 4.10 (when approved by the Board) to protect indigenous peoples' rights. Within that framework, special emphasis will be given to support policy dialogue aimed at legislative reforms that will protect the forest land ownership and access rights of the poorest
- Supporting collaborative forest management (CFM), improving forest harvesting and management programs, and incorporating safeguard measures aimed at minimizing the risk of more powerful members of the community or outside commercial interests appropriating nontimber forest products on which the poorest depend
- Through a combination of ESW, policy dialogue, and policy-based investment programs, tackling difficult issues such as how to mitigate undesirable impacts of globalization and market liberalization in the forest and other industrial sectors on poor forest-dwelling societies. These impacts include expanding commercial-scale forest and agribusiness operations, mining, oil exploration operations that may lead to appropriation, or privatization of forest resources that the very poor continue to need to access as common property

Sharing by the state and the rural poor in forest benefits. One of the more fundamental policy issues that the Bank will need to address is competition between smallholder and community producers for their shares of the economic benefits from forest outputs. Governments impose taxes and other charges and costs on forest outputs. In addition, state forest agencies often support local producers under one program and subsidize state-run operations, or large corporate activities that compete unfairly with local producers, under another. In the short term, the scope for improving the situation probably lies primarily in removing or relaxing regulatory provisions that reinforce the structural and scale advantages that the state possesses as a producer of many forest products. The relationship between forest departments and small local forest users also could be improved by separating governmental forest departments' regulatory functions from delivery of forest support services.

In the longer term, a logical solution in some situations will be to phase out state production in markets in which smallholder production has a comparative advantage. This shift also would contribute to meet a more fundamental concern that has been raised: the potential for the rural poor to benefit will continue to be limited so long as they are unable to participate in the more profitable and dynamic production activities.

Integrating forest activities focused on the very poor in rural development strategies. Implementation of the revised Forest Strategy will make a significant contribution to meeting the goals of the Bank's revised Rural Development Strategy. This strategy refocuses the rural development process to concentrate on improving the well-being of rural people and reducing poverty in the widest sense. The latter entails much more than increasing the average income of rural people. It envisions improving the quality of rural life. The underlying concept for this strategy is a developing world in which

- Rural residents enjoy a standard of living and a quality of life that is not significantly below that available to urban residents;
- Rural communities offer equitable economic opportunities for all their residents regardless of income, status, or gender;

- Rural communities become vibrant, sustainable, and attractive places to live and work;
- Rural areas contribute to national development and the overall economy and are dynamically linked to urban areas;
- Rural areas adapt to ongoing economic, social, cultural, economic, and technological change.

Special attention will be given to incorporate sustainable forest resource management and agroforestry initiatives into Bank-supported dry land, water catchments, wasteland reclamation, and other rural development programs. Past Bank experience in Burkina Faso (community-based natural resources management programs), Turkey (Anatolia Watershed Project), and China (poverty reduction–oriented projects) provides useful lessons of wider relevance.

Regarding the interests of the poorest of the poor, the arguments in favor of joint and collaborative management have become more prominent as it has become apparent that user communities and institutions often are unable to take responsibility for control and management unaided. Governments also have favored joint management, because transferring management and protection responsibilities to the community level can help offset reduced budgetary resources available to forest departments. However, such forms of local management run the risk of the state's continuing to exert too much control.

While, in general, Bank experiences with collaborative management have been reported as successful and have resulted in a greater degree of involvement of rural users, this involvement has not always benefited the poorest of the poor. Some earlier interventions were based on insufficient understanding of the circumstances under which collective management is appropriate and of the realities of the rural populations involved. Experience also has exposed serious problems in the ways in which governments have devolved responsibilities for forest management. Accordingly, the Bank's strategy will seek to address through ESW and policy dialogue some of the inherent weaknesses in current CFM approaches. These weaknesses include failure by governments to transfer effective authority, restrictions on rights granted to the poor, and ineffective and frequently inequitable local institutions.

Emphasis will be given to situations in which land, labor, and capital availability favor multistory "home gardens," which increase land productivity. In such situations, agroforestry systems as demonstrated by Bank-supported International Centre for Research on Agroforestry (ICRAF) research can significantly reduce poverty by providing low-cost inputs from, for example, nitrogen-fixing fodder trees that reduce the dependency of poorer farmers on purchased fertilizer or animal feed.

Fostering linkages between the forest industry and rural poor. The Bank will foster partnerships between industry and local communities and smallholders to provide access to credit, extension, markets, and skills development. It also will seek to strengthen cooperative arrangements to improve the bargaining position of the rural poor and enable them to participate in the benefits from downstream processing and trading. These improvements could include removing government subsidies to expanding industrial capacity that impinges on small-scale producers—as happened with rattan production in Indonesia.

An underlying objective of Bank strategy in this area will be to encourage situations in which the bulk of timber or industrial roundwood requirements of larger forest industries such as pulp and paper are met from common property, or collaboratively managed or privately owned forest resources. For example, fostering collaboration between small, private woodland owners and industrial companies is a major focus of ongoing Bank-supported strategies in Bulgaria, Romania, and several other transition countries of Eastern Europe.

Tree out-grower arrangements and other forms of contractual agreements such as those fostered by Bank loans in Argentina, Brazil, and the Philippines can provide important links to markets and support for farmers, but care needs to be taken to ensure that they are targeted at those able to benefit from them. Growing can be especially appropriate for smallholders who have sufficient annual income from other sources to secure their ongoing needs and land they can use for trees that is not needed for food crop production or for other more basic needs. Tree growing is likely to be an attractive option in these circumstances in which the features of an assured market and access to technical advice and inputs make tree crops a more stable source of income than alternative uses of the land. These features and the probable need to have title to their land to be eligible for a loan indicate that tree out-growing is unlikely to be feasible for many smaller or extremely poor farming families. It is more likely to be an appropriate activity for the moderately poor.

INTEGRATING FORESTS IN SUSTAINABLE ECONOMIC DEVELOPMENT

The element, integrating forests in sustainable economic development, focuses on capturing the large current and potential asset value of forests (outlined in appendix 3) and proposes that this asset needs to be well managed to yield sustainable economic growth. Good management of the forest resource includes controlling illegal activities, promoting effective participation, and managing strong cross-sectoral and macroeconomic linkages with forests.

Cross-Sectoral and Macroeconomic Linkages

The Bank must have an appreciation of how its actions and investments in other sectors, or at the macroeconomic level, will impact forests and forest peoples. It then must act on this information to incorporate measures to offset or minimize the impacts. At present, in the Bank, the level of integration of major forest concerns in the design of larger adjustment and cross-sectoral activities is relatively low. The priorities for Bank engagement in this area will be as follows:

Cross-sectoral approaches. Policies and projects need to be analyzed and coordinated to ensure a cross-sectoral approach to planning and implementation of SFM and forest conservation and development. The Bank's strategy will give special emphasis to supporting the large number of rural poor living within forest margins or outside forests (predominantly agricultural populations) who are able to access forests, tree stocks outside forests, and trees on farms, and to respond to market opportunities. Forestry assistance will be defined broadly to encompass all tree stocks and activities on which they are based.

The Bank will invest more in analyzing the potential impacts on forests of large programs in rural development and infrastructure.[2] In this strategy, a major task for Bank operations staff and management will be to develop the necessary linkages and processes to ensure that when large rural programs are under consideration, forest issues are factored in. Most of the poor who live in or near forests are associated with some form of agriculture and are significantly dependent on nearby forests for aspects of their livelihood. The broad patterns of development that occur should preserve this balance (appendix 3). Closer linkages will need to be developed with agencies such as ICRAF and the Center for International Forest Research (CIFOR), which have considerable experience in designing and implementing these options.

The Bank will seek opportunities to support collaborative forest management, in particular, building on past Bank experiences in countries such as India, Mali, Mexico, and Niger. Much has been learned in these countries about how to involve user groups in sustainably harvesting and managing dry land savanna fuelwood resources.

A key concern is to prevent any significant conversion or degradation damage to critical habitat area, including all critical forest areas. The fundamental guidance given to Bank staff and clients is that in assessing the significance of change they must take a precautionary approach (see OP 4:04, Natural Habitats, paragraph 1).

In assessing the significance of any change likely to be induced by a particular investment operation, key considerations include the intensity and scale of the operation. Furthermore, the acceptability of any change will be determined by the environmental, social, and managerial context of the forest or habitat area concerned. In some countries, official guidance suggests that loss or inability to reproduce of 1 percent of any species would be considered "significant." However, this does not take into consideration the variation between species in terms of reproduction rates, or the ability of the population to recover following impacts (Treweek 1999). In some park and wildland management settings, systematic planning frameworks for assessing the acceptability of change are available (see McCool and Cole 1997).

This issue will be addressed in greater detail, and in the form of guidelines for determining significance in a specific context, in the forthcoming Sourcebook for implementation of this strategy, and consideration will also be given to whether any revision of existing definitions of "significant" need to be amended in various OPs.

Macroeconomic impacts. Relatively little analysis has been done in the Bank on the impact of macroeconomic policies on forest outcomes. Furthermore, most of what has been done directly addresses the impacts of specific provisions for forests and forestry that occasionally have been included in these larger policy frameworks. The current revision of the Bank's OD on adjustment operations will consider such impacts and issues. In the meantime, careful attention will be paid to policies with the potential to have major impacts on forests, including trade and tariff policies favoring extensive agriculture or tree crops and overall public expenditure reduction (appendix 3).

During the consultative phase in formulation of this strategy, a number of commentators noted that the proposed new forestry policy did not address potential forestry impacts of programs supported by Bank adjustment lending. Because the time frame for an update of the Bank's adjustment lending policy remained uncertain, some suggested that the Bank put in place a transparent set of procedures for systematically identifying significant forestry impacts associated with Bank adjustment operations, analyzing such impacts, and, if necessary, adopting and implementing appropriate mitigating measures.

In response to these concerns, the Bank has developed the following approach to deal with development of a new adjustment lending OP in a timely manner and to address potential problems in the intervening period:

- The Bank plans to address the treatment of possible forestry impacts of programs supported by Bank adjustment operations as part of the treatment of overall environmental impacts of such programs in the ongoing update of OD 8.60 on adjustment lending into a new OP/BP8.60. This approach will put the forestry impacts of Bank-supported reform programs into the appropriate context of the full reform program, while avoiding a fragmentation of Bank OP on adjustment lending into a multitude of sector-specific provisions.
- At the time of writing of this strategy, an issues paper on this new OP had been posted for public comment and review. An initial draft of the OP itself had been prepared on the basis of responses and specific consultations and was presented to the Committee on Development Effectiveness of the Board of Executive Directors. A new version is currently under preparation. Public consultation meetings were held in London, Washington, D.C., and Amman, Jordan. Further meetings are scheduled in Dakar, Dar es Salaam, Mexico City, Seoul, Warsaw, and South Asia.
- For the period until the new OP/BP8.60 is in place, the Bank intends to put in place a set of transparent arrangements for the treatment of forestry impacts associated with Bank-supported adjustment operations, including systematically identifying possible significant forestry impacts associated with Bank adjustment operations, analyzing such impacts, and, if necessary, adopting and implementing appropriate mitigating measures. A three-pronged approach is planned, beginning with pilots focused on cases with clear forest linkages that may provide useful lessons with broader applicability.

1. The Bank will continue implementing the good practice advice in the current OP on adjustment. This good practice is summarized in the Operational Memorandum "Clarification of Current Bank Policy on Adjustment Lending" of June 5, 2000: ". . . it is good practice for Bank staff, in preparing appropriate assistance programs, to review environmental policies and practices in the country, take account of any relevant findings and recommendations of such reviews in the design of structural adjustment programs, and identify the linkages between the various reforms proposed and the environment. If there are negative linkages, it is good practice to devise specific measures to counteract the possible negative effects, or explain how mitigation is being achieved elsewhere within the Bank's Country Assistance Strategy."

2. Building on this good practice, the Bank plans to put in place a process whereby Regional Vice Presidencies will work with ESSD and Operations Policy and Country Services (OPCS) to screen forthcoming adjustment operations—as early as possible during program preparation—for specific policies supported by the operation with possible forestry impacts. In those cases where impacts are likely to be significant, the Bank envisages a more detailed follow-on review with a view to assisting governments in developing measures to help avoid/mitigate them.

3. In assessing potential adverse impacts on forests, the Bank expects to draw on a range of technical expertise and country knowledge. The Bank is considering calling on the assistance of the TAG, a multistakeholder group that was set up to assist the Bank in developing its Forest Policy and Strategy. The TAG could act as a roster of expertise, from which specific individuals or groups would be drawn to assist the Bank, on a case-by-case basis at the request of the Bank, by providing input in judging forest impacts of policy reform and possible mitigation measures.

Expanding Nonfarm Rural Activities: Role of Small-Scale Forest Product Enterprises

In nearly every country in which such information exists, small-scale forest product activities are among the three largest categories of nonfarm rural commercial activity in terms of numbers of people engaged in these activities. Bank interventions in this area have the potential to benefit large numbers of people.

The Bank will target its assistance to small, wood-based enterprises to help the rural poor successfully engage in processing and trading. The Bank will support

- Market research and improved marketing strategies;
- Provision of micro credit (as a component of more broad-based Bank support for small-scale enterprise development);
- Technical and business management training;
- ESW aimed at identifying regulatory and other constraints to small-scale wood enterprises.

Governance in the Forest Sector: Forest Crime, Corruption, and Regulation

Examples in appendix 3 show the costs and the pervasiveness of crime and corruption in the forest sectors of many countries. Persistently low resource rents and nontransparent resource allocation procedures are strong indicators of the potential for major forest corruption and governance problems. Additional indicators include market cartels and other distortions; inadequate forest conservation and protection measures; significant and organized illegal activity including timber theft, misrepresentation of volumes, species, or the quality of log harvests; poaching; and commercial-scale encroachment of agricultural or other activities on lands designated as permanent forest zones. Poor governance also usually is reflected in unclear and conflicting tenure or management mandates for forest lands and the systematic exclusion of local peoples from decision-making that affects forest lands.

Illegal logging and corruption. By far, the greatest forest concern for many governments is the cost of illegal logging and forest-related corruption. In this area, the World Bank will take these actions:

- Use its involvement in formulating CASs and NFPs (chapter 3) to support legislative and institutional reforms that will help make forest law enforcement more effective and just. The Bank is developing guidelines for its future involvement in forest law enforcement activities
- Finance monitoring programs that support SFM planning and help identify the extent of problems. Along with deeper analysis on the underlying incentives to forest users and managers, these programs will provide valuable input to governments as they develop NFPs or strategies
- Initiate dialogue with countries and regions to foster collaborative partnerships among national governments, local people living in and near forest areas, private industry, NGOs, donors, and other stakeholders to more effectively monitor and control forest use. These partnerships will include a wide range of measures aimed at detecting, preventing, and suppressing illegal forest activities. The partnerships will give particular emphasis to strengthening civil society and key actors such as the judiciary, law enforcement branches, and government forest management agencies. The Bank also will initiate dialogue with other intergovernmental organizations, such as the FAO and International Tropical Timber Organization (ITTO), and international environmental NGOs, such as Global Witness and Global Forest Watch, which work in this area

Although experience in this area is limited, governments, including in Cambodia and Indonesia, already have given encouraging responses to Bank-supported forest law enforcement initiatives. As a result, similar activities are being planned in other regions.

Reforming timber concession policies. The Bank will promote the use of rigorously designed regulatory frameworks for timber concessions to enhance the contribution of planned forest use to economic and social development and environmental protection. Forest concessions in both tropical and temperate countries often have not yielded an equitable distribution of economic benefits. In addition, intensive exploitation of forests frequently has led to the degradation of the forest resource and the unnecessary loss of both timber and nontimber values.

Policy changes will have to be coupled with addressing the technical competency of forest officials, eliminating corruption in field-level operations, and involving local communities and their representatives in forest management, including the concession allocation processes. The Bank will encourage governments to take advantage of growing opportunities to engage independent third-party certification bodies in performance-based monitoring of forest harvesting and management operations.

Building a Role for Civil Society in Sustainable Forest Management

Sustainable forest management issue. Recognition is growing that effective forest management is critical to sustainable development, particularly for local or national economies that are based significantly on the use of forest resources. Given the extent of the resource in many countries, few governments or community stakeholders accept that logging should be banned in all accessible forests. Therefore, the key questions in most countries are where logging should occur and how well it will be managed.

In a land-use planning context, the major challenge is to ensure that extractive activities such as logging take place in areas in which the benefits of the activities outweigh any social or environmental costs that they may engender. A priority here is to ensure that areas of special conservation value or social significance are conserved in culturally appropriate forms of Protected Areas. However, few countries have been able to afford to allocate more than 10 to 20 percent of their forest areas for this purpose. Thus, there is a growing consensus among ecologists and resource management specialists that biodiversity cannot be adequately conserved by Protected Areas alone.

This reality has proved to be the case in North America, which arguably has the world's best-funded and most extensive national park–wilderness conservation system. The capacity of Protected Areas to protect biodiversity is likely to be even less in tropical areas with their higher levels of habitat differentiation, disjunctive species distributions, low aerial population densities of many species, and high endemism. In addition, IUCN global data indicate that some 85 percent of existing Protected Areas have human populations living either inside or immediately adjacent to the reserve. Therefore, the long-term future of biodiversity will depend just as much on sympathetic management of productive, humanized landscapes as it will on the balanced selection and management of traditional Protected Areas.

A key part of this process, which will need to be applied on a larger portion of the world's accessible natural forests, is to combine conservation and production goals in the same area. Such an approach needs to be based on consensus. In addition, wherever ownership and tenure arrangements permit, planning and management should be collaborative. Ecological and silvicultural knowledge is incomplete. Nevertheless, sufficient knowledge is available and proven for all major forest types, including tropical humid forests, for natural forests to be managed in ways that will maintain high levels of environmental, economic, and social value. However, even this available knowledge frequently is not put into practice because of a lack of appropriate incentives for forest users and inefficient monitoring by forest services. Larger economic and governance issues often are at the root of these problems.

What the Bank should and should not do in forests stimulated considerable controversy, both inside and outside the Bank, while and after the 1991 strategy and the 1993 policy were developed. These issues remain contentious.

A complicating factor has been the continuing, still unresolved, and, at times, highly emotional debate over definitions of SFM. It is a given that many national governments of forest-rich tropical countries, such as Brazil, Cameroon, and Indonesia, will continue to sanction harvesting operations in significant areas of their primary tropical forests. Therefore, the regional consultations leading up to formulation of this strategy voiced considerable agreement for the Bank to become actively involved in promoting more sustainable forms of use in forests outside formally Protected Areas. Support for this type of activity was seen as essential if the Bank is to be a meaningful actor in the struggle to contain socially, ecologically, and economically damaging forest activities that reduce forests' potential to contribute to sustainable poverty reduction.

As an interim step, pending a clearer international consensus on this issue, the Bank has agreed with leading international conservation agencies that it will encourage the widespread use of internationally agreed criteria and indicators for SFM. These criteria include those defined by the ITTC, discussed in the Intergovernmental Panel on Forests (IPF), Intergovernmental Forum on Forests (IFF), and embodied in the principles and criteria of bodies such as the Forest Stewardship Council (FSC). The Bank worked with WWF and a wide range of forest sector stakeholders to design a set of principles and criteria for certification systems to move forest management toward SFM. These principles and criteria are set out in the World Bank/WWF Alliance 1998 "Guidance Note for Improved Forest

Management and Certification Target: Achieving the Independent Certification of 200 Million Hectares of Well Managed Production Forests by the Year 2005."[3] It would be divisive and potentially misleading to suggest that SFM can be defined unambiguously and then assumed to be always attainable within the time frame of a given intervention. Rather, this strategy will emphasize improving forest management along the lines of the pyramid approach described below. This approach has been developed by the World Bank/WWF Alliance to support the Alliance's global target for SFM: to achieve 200 million ha of independently certified production forests by 2005.

A number of specific issues arise in relation to the Bank's involvement in certification:

- Under the principles and criteria set out above, the Bank has accepted the principle of independent monitoring of forest operations. However, the Bank has not endorsed any particular certification system but will assess particular approaches in relation to their compliance with these principles and criteria. The Bank recognizes the ongoing "mutual recognition" debate in the international community to harmonize acceptable standards and approaches and expects that these principles and criteria will contribute to these discussions.
- There is some debate on the issue of whether a fixed time limit should be applied to achievement of certification. The new OP will require the negotiation and disclosure of a time-bound action plan acceptable to the Bank wherever support is given to improving operations that cannot currently meet the requirements of an acceptable certification system. Because of the wide range of circumstances likely to be encountered in different enterprises and in different countries, selection of an arbitrary time limit for all cases would not be advisable. Timing should be determined on a case-by-case basis during project preparation with full disclosure. The forthcoming Sourcebook will address the subject of how an acceptable time frame for achievement of certification standard operations should be determined in each individual case.
- There is an issue of how firms suitable for certification should be selected; in particular, the weight that should be given to a proven track record in certification when selecting firms. It would not be advisable to insist that any firm wishing to compete for a certification contract should necessarily have a track record, because this would exclude new firms from entry, even if they had clear capability and access to experienced staff. It might also result in a virtual monopoly for some existing certification firms or agencies in some locations, and the Bank has consistently tried to avoid this outcome by ensuring that objective standards of performance and process became the criteria for evaluation, not mere existence. This also relates to the first issue raised above—the need for the Bank to avoid endorsing a particular firm or agency. There is also the question of country capacity building or ownership. If client countries or regions develop systems that clearly meet all the performance standards the Bank requires, then giving a priori mandatory preference to established outside bodies/systems would potentially inhibit country ownership for good forest practice and the transparency of robust certification processes. However, selection of firms to undertake certification under Bank-financed projects will be subject to standard due diligence provisions under the procurement guidelines, requiring that assessment of capability of firms to be included for consideration is carried out. More generally, the forthcoming Sourcebook will provide criteria and guidelines for assessing certification firms, and it is evident that these will require that due weight be given to track record when formulating a choice.
- In projects involving certification in which the Bank is investing, it will need to take responsibility for decisions on the criteria to be employed in certification, and also for ensuring that these criteria are then applied in the field appropriately. This will be supported by material in the Sourcebook and in other material on what the standards and norms are, derived from the growing international literature on this subject. Moreover, it is almost certain that any project in which the Bank was intending to be involved with forest operations and certification will be assigned Environmental Category A. This means it will require an advisory panel, as required in the Environmental Assessment OP (OP 4.01), and in cases involving certification this panel would include expertise on certification, and thus would allow for independent evaluation of the scheme being considered.
- The question of adding chain-of-custody requirements to certification has been raised. Chain of custody is a valid and useful activity, which serves the purpose of verifying the origins of wood and

allows for control over illegal logging operations. The Bank has used chain-of-custody log tracking in places such as Papua New Guinea, where underreporting and other corrupt practices were issues and where it was possible to attain coverage at all export points. However, the need for this, and its practicability, should be assessed on a case-by-case basis; it will not always be possible to attach chain-of-custody activities. One factor in the decision will be the extent of coverage of all forest operations intended under the investment being considered. Certification can be applied on a concession or field operation level, and will address the issue of illegal removals from such operations. Chain of custody relies heavily on sufficient coverage to be able to stem the tide of illegal timber in aggregate, and therefore demands heavy and broad investment to achieve. Guidelines on the decision to apply chain of custody will be included in the forthcoming Sourcebook.

The Bank believes that, in many countries, greater involvement of local communities and other key stakeholders in forest management and planning is needed to produce equitable outcomes and raise the overall social value of forests. In some cases, the need for greater transparency and accountability at the local level will require the use of local stakeholder assessment as an alternative to third-party assessment of commercial-scale operations.

The Bank will encourage national governments to develop standards for natural forest management and forest restoration that are locally relevant and meet internationally recognized principles and criteria for SFM. The institution also will provide support to national governments to create representative, multistakeholder, and independent forest monitoring bodies.

The "Pyramid"—a diagnostic and planning tool for good forest governance. Forest management, for both exploitation and conservation objectives, depends critically on matters far from the forest itself. It depends on the extent and quality of the *enabling* policy, legal, and institutional conditions—on good *forest governance.* In an effort to introduce a simple, but robust, means for stakeholders to work together in assessing, and in planning, these key enabling conditions, the World Bank introduces the "Pyramid" (figure 2.1) which was developed in the framework of the World Bank/WWF Forest Alliance. The objective of the Pyramid is to offer a framework to stimulate participatory assessment and target setting in forest governance at the country level. The concept behind this analytical framework is that some elements of good forest governance are common to a wide range of nations. By grouping these elements in several tiers of complexity, the Pyramid can serve as a country-level planning tool in forest management. In a multistakeholder process, the tool can be used to assess the status of forest governance using a scoring system to identify what is working, what is missing, and what needs to be done for different elements of forest governance. The tool is intended to fill the "forest governance gap" between assessing and accelerating field-level progress in SFM and international policy, assessment, and reporting. By filling this gap, stakeholders' capabilities to deliver national governance that supports local forest governance—and potentially improves international forest governance—can be improved.

To assess the governance situation in Bank client countries prior to a forest operation, Bank task managers are interested in using this tool to analyze the elements of forest governance and in prioritizing strategic areas for cooperation with these countries on forest sector issues. Piloting countries include Kazakhstan, Armenia, and Cambodia.

The individual tiers of the Pyramid are seen as mutually reinforcing and not necessarily so rigidly linked as to form preconditions. The Pyramid concept acknowledges different stages and approaches in building improved forest management and avoids forcing the pace with unsustainable solutions beyond the absorptive capacities of the governments and communities concerned. It also helps identify milestones that acknowledge incremental gains in places in which forest governance problems are greatest. In addition, it helps encourage policy dialogue aimed at eliminating those constraints.

Approach to World Bank's Forest Policy and Sustainable Forest Management

To respond to OED's recommendations, the concerns raised by many in the external and internal consultations, and the analytical studies on advances in understanding forest issues, it was necessary to modify the 1993 Forest Policy, OP 4.36, in both scope and clarity. OP 4.36 also needed revision to be consistent with this revised strategy.

Since the 1992 Rio Earth Summit, the international community has agreed that the international forest dialogue must transcend discussion solely of

FIGURE 2.1
The "Pyramid"—A Diagnostic and Planning Tool for Good Forest Governance

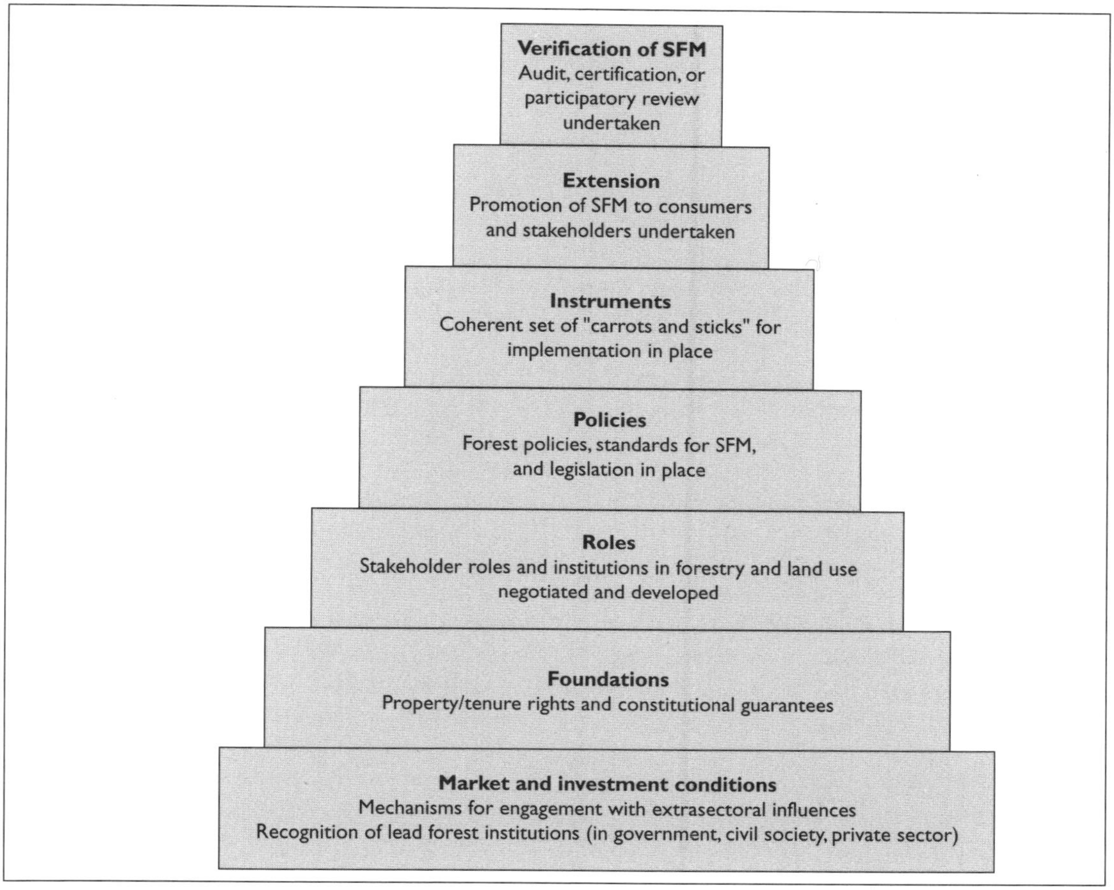

Source: Mayers, Bass, and Macqueen 2002.

tropical forests to focus on all forests in all countries. In addition, especially since the advent of the Bank's major engagement with the Eastern European and Central Asian countries, temperate or boreal forests have become major elements in the Bank's forest lending program. However, OP 4.36 focused primarily on tropical moist forests. In addition, OP 4.36 was issued before the approach to independent assessment of the quality of forest operations on the basis of performance-based certification was developed. Independent certification is becoming the industrial standard for improved forest management. OP 4.36 was silent on certification.

Furthermore, and as demonstrated by OED, OP 4.36 had a "chilling effect" on Bank Group involvement in virtually all aspects of tropical forest management. This reduced Bank involvement created a particular difficulty in relation to the major targets for the achievement of SFM that the Bank's president presented to the Special Rio+5 Session of the General Assembly of the United Nations in 1997 and that the Bank and WWF, its partner in the Global Alliance for Forest Conservation and Sustainable Use, have adopted as one of the major objectives of that partnership.

As noted elsewhere in this paper, the target of 200 million ha under SFM by 2005 will be difficult to achieve in any event—especially given the restrictions OP 4.36 has applied to some of the activities necessary to achieve the target. The longer the blanket restriction on financing logging operations in tropical moist forests under any circumstances remained in place, the more difficult this sustainability goal became.[4] This prohibition on Bank involvement in financing commercial harvesting "in any circumstances" was justified at the time it was adopted given that the vast majority of tropical forest harvesting practices were demonstrably unsustainable.[5] However, in the current situation, a viable policy must allow for circumstances in which such involvement demonstrably would improve conservation and development outcomes. OP 4.36, as it was, inhibited the Bank, particularly the IFC, from assisting countries and the private sector to improve their forest operations or, for that matter, from engaging in a meaningful dialogue on SFM. In consultations on this subject, several client countries indicated their reluctance to engage with the Bank in this area when the Bank takes such a prohibitive stance on financing, regardless of the quality of the forest operation.

The new forest OP redresses this weakness and enables the Bank to more proactively promote desirable forest conservation and development outcomes. The new policy should prevent the financing of commercial logging operations in critical forests in *all* major forest types, not just primary tropical moist forest. It also ensures that no Bank financing will be provided to logging operations, in any type of forest, in which these operations do not meet acceptable standards of sustainability, as verified through certification processes.

The new policy will help the Bank to achieve the major objectives it has set in this strategy, by providing a clear operational framework based on rigorous, performance-based independent certification and consistent application of the Bank's overall safeguards policy framework. It will signal to the forest community that the Bank is supportive of well-managed forest operations that conform to recognized international certification standards.

Risks and issues. The new, more proactive policy that enables the Bank to become involved in the sustainable management of production forests carries some risk. However, there also were major reputational risks for the Bank in retaining the previous policy, which did not explicitly address the issue of the temperate forests at all. The policy was criticized severely by many stakeholders and interested parties outside the Bank as being a major factor in the Bank's failure to make a significant contribution to forest conservation and development. Critics also saw the 1993 policy as failing to meet the challenge of mobilizing the considerable potential contribution of forest conservation in the attack on poverty and economic and social underdevelopment.

A further implication of adopting the new policy approach proposed above is that it will require monitoring of all prospective and ongoing operations for compliance with these standards.

Major underlying arguments for the new policy approach were, first, that the Bank needs to recognize in its policy the reality that many forests around the world that are accessible and commercially valuable *will* be used for timber production—sooner, in all likelihood, than later. Under the new policy, the Bank will commit to avoid at all costs financing commercial logging in critical natural habitats and other nonsustainable forms of commercial-scale logging. However, it will be prepared to use its influence and financial resources to help improve the quality and sustainability of commercial operations outside these critical forest areas and ensure that a more equitable share of the benefits from commercial activities is directed toward local communities living in or near forest areas. Very often, the real choice available in this sector is not between logging and doing something less invasive and damaging to the forest, especially when these alternatives have not been developed to large-scale market viability. Rather, it is between doing logging reasonably well, or doing it very destructively so that conversion of the logged-over site to other uses becomes almost inevitable. In addition, as noted earlier, the Bank will require all projects involving the utilization of forest resources to include an evaluation of the prospects for new markets and marketing arrangements for all the goods and services that can be produced from well-managed forests. This will help to ensure better market-based incentives for good forest practice and more sustainable and conservation-oriented patterns of forest land use.

Second, as discussed earlier in this chapter, since the previous Forest Policy was adopted in the early 1990s, powerful citizen-driven certification and independent monitoring procedures have emerged. These processes are helping to ensure that, increasingly, large commercial-scale logging and forest industrial operations around the world are being subjected to and accepting close scrutiny and systematic assessment of their impact on both people and forests. In partnership with the WWF, the Bank has committed to a global target of 200 million ha of independently certified, well-managed forest by 2005. This fact should help to allay the fears of some environmentally concerned groups that, by revising its OP on forests, the Bank would merely be opening the flood gates to a massive expansion of nonsustainable logging.

Third, recognition is growing around the world of the complementarity of forest conservation and sustainable forest-use strategies at the landscape level.

Governments of leading forested countries such as Brazil, Canada, and Malaysia have adopted forest land-use "zoning" strategies that allow for a continuum of (a) totally Protected Areas; (b) multiple-use forests in which overriding concerns for protection of biodiversity and other environmental benefits are integrated with continued harvesting of those numerous forest products that are essential for subsistence and local use; and (c) production forests in which independently certified, sustainable timber harvesting embraces the goals of preservation of vital forest environmental functions.[6] Such an integrated, ecosystem-based approach drives the mix of conservation and sustainable management objectives defined in the World Bank/WWF Alliance for Forest Conservation and Sustainable Use. The new Forest Policy enables the Bank to become a more proactive partner in addressing the urgent need to assist client countries to achieve better conservation and development outcomes with their forest resources.

PROTECTING VITAL LOCAL AND GLOBAL FOREST VALUES

The third major pillar of the revised strategy is to protect vital global and local environmental services and values that come from forests. This pillar focuses on the problem that domestic and international markets and individual incentives do not account for these externalities, despite their importance for sustaining rural incomes, economic growth, and ecosystem protection. There also is insufficient international financing to offset these tendencies at the national level.

Evolving Perception of Protected Area Strategies

One of the problems identified by OED is the difficulty inherent in protecting forests in high demand for a range of—frequently, mutually exclusive—uses by competing groups within society. In cooperation with the GEF, the Bank has generated significant increases in funding for biodiversity protection and related purposes (chapter 3 and appendix 4). Nevertheless, the scale of this contribution is dwarfed by the incursions being made into forest areas—including forests that ideally should be protected from invasive use. Moreover, Bank client governments do not, by and large, wish to borrow funds for forest protection. The reality, therefore, is that, unless significant additional funds at highly concessional or grant terms blended from multiple sources can be made available for protection, or effective markets for the ecosystem values of forests developed, the problem is likely to worsen. It is important to bear these facts in mind when considering the Bank strategy in this area.

This larger funding problem is not always apparent when protection activities are being considered. Unlike the Alliance sustainability target discussed above, achievement of the Alliance 2005 target of 50 million ha of new Protected Areas is well on the way to realization. Through the Critical Ecosystem Partnership Fund, another major effort is under way to set aside important biodiversity hot spots. However, if these efforts are to lead to protection across the board in remaining natural forests, and not only in selected areas, perceptions of Protected Areas that would give high priority to setting aside discrete wilderness areas and biodiversity reserves and excluding them from all forms of human use will have to evolve. There are signs that this change in perception is happening. It is now widely recognized that local communities and forest-fringe farmers can play a key role in biodiversity preservation. There is a trend toward a wider definition of Protected Areas that embraces the concepts of IUCN Category VI: "areas containing predominantly natural systems managed to ensure long-term protection and maintenance of biological diversity while providing at the same time a sustainable flow of natural products and services to meet community needs."

Adopting and promoting the expansion of IUCN Category VI Protected Areas have major implications for the Bank's revised strategy. These changes suggest the possibility that large areas of community-managed forest, such as those in India and many other parts of the world, also will be recognized as making major contributions to the preservation of biodiversity, carbon sequestration, and other environmental services.

Fostering Markets for Ecological Public Goods

A second key element for the Bank in expanding the potential to protect forests, discussed further in chapter 3, is helping to facilitate and build new markets for forest ecosystem services. These services include

- Services of local and national significance, including watershed, soil, and other environmental sustainability purposes;

- Global forest common goods, including biodiversity—already the subject of debt-for-nature swaps—and conservation concessions;
- Carbon, potentially a much larger and more lucrative source of funding for the retention of forests.

Developing such markets is an important factor in enabling countries to recognize, and then realize, the real value of their forests. The increased revenues and incomes that national governments and local communities can reap from these environmental services can serve as major incentives to sustain forests—rather than to cut or burn them for short-term returns. However, these long-term uses will become attractive to national decisionmakers only when reliable market mechanisms are in place.

The two major opportunities in this area that the Bank will pursue are:

- Helping to build and finance markets for international public goods, such as carbon and biodiversity. Major potential appears to exist for use of the guarantee instrument. In September 2000, the Bank approved an experimental Partial Risk Guarantee to encourage international private investment in Russia's forest products sector through a guarantee arrangement between the government and the Bank. This instrument will underwrite the risk that adverse and unpredictable policy changes pose for private investors in the Russian economic environment. In this case, it is a potentially powerful approach for building the partnership between the Bank and the private sector. In other circumstances, it could be an equally powerful approach to underwriting commitment to sequestration or protection zones, when combined with grant-based financing for these purposes
- Taking advantage of a considerably underused opportunity by assisting governments to design, implement, and finance the start-up of effective national markets for environmental services provided by forests. A number of options have been explored, by the Bank and others, including financing regulation, environmental services payments, tradable rights development systems, disincentive systems, and financing innovation (World Bank 2001c). All of these approaches have been applied successfully to protect or preserve aspects of the forest ecosystem that would otherwise be in jeopardy

Fostering the Linkage between Forests and Climate Change

Climate change is projected to significantly alter the composition and possibly the productivity of all types of forests, but particularly in the tropical and boreal latitudes. Such alterations are due to changes in the mean, variability, and extremes of temperature and precipitation, coupled with an increase in disturbance regimes, that is, pest outbreaks and fires. Nevertheless, forests and forestry can play a major role in mitigating greenhouse gas emissions and in preparing client countries to adapt to today's climate variability and long-term climate change.[7,8]

The economic consequences of climate change are expected to fall disproportionately on developing countries and, within these countries, on the poor. Climate change will have its greatest negative impact in the tropics and subtropics, with higher temperatures, increases and decreases in total rainfall, more heavy precipitation events, more El Nino–like conditions, and an increase in tropical cyclone wind and precipitation intensities. These changes are likely to lead to, among other things, an increase in floods, droughts, and landslides, which often affect the poor most of all and exacerbate their already inadequate living conditions. Climate change is expected to exacerbate inequities in health status and access to adequate food, clean water, and other resources. Although some adaptation is possible, developing countries, especially the poor within developing countries, are unlikely to have the institutional, technical, and financial means to adapt.

The role of forests in reducing and preventing risks and vulnerability from natural disasters at the local level is well known and does not require a new agenda. For example, reforestation can help avoid landslides, avalanches, and flash floods in headwater streams. Forest products and tree crops are less vulnerable to weather hazards and often provide a cushion for the poor in times of food shortage. In the present strategy, adaptation measures involving forestry are considered an integral part of the proposed poverty reduction strategy.

Of a more controversial nature is the use of forests' capacity to sequester carbon and hold it over long periods. While the bulk of global carbon emissions come from fossil fuel consumption in industrialized countries, deforestation and subsequent land-use change also are significant, especially in the tropics. In the tropics, forest loss causes between 10 and 30 percent of the global CO_2 releases.

Afforestation and reforestation are regarded as efficient means to sequester atmospheric carbon. Moreover, conserving and sustainably managing existing natural forests and forest soils, which are large stores of carbon, can significantly reduce greenhouse gas emissions. Arrangements for the first commitment period (ending 2012) under the Kyoto Protocol allow afforestation and reforestation but not deforestation to be eligible activities under Article 12 of the Kyoto Protocol, that is, the CDM.[9] These allowances, and their potential longer-term extension to areas such as avoided deforestation, have the potential to provide new revenue streams to local economies for forest protection and forest management services. Industrialized countries, including those with transitional economies, already are allowed to meet their obligations by trading carbon using afforestation, reforestation, and avoided deforestation activities under Article 6, Joint Implementation. Hence, forests could have a major role in climate mitigation in the near future, and the Bank could take a leadership role in developing and facilitating equitable market arrangements that will enable this to happen.

The Intergovernmental Panel on Climate Change special report, "Land Use, Land-Use Change, and Forestry" (2000), specifies the role of forests and forestry within the climate change framework. Forestry projects are likely to be as, or more, cost-effective in offsetting carbon emissions than other options and could have a significant impact on sustainable development in many of the Bank's client countries. Taking into account the considerable political uncertainties and controversy that prevail, the Bank needs to consider the perceived value of forests offered within the framework of the United Nations Convention on Climate Change. As stated earlier, some forestry activities are already eligible under the CDM, and their eligibility will stimulate the development of international markets for carbon credits from forestry through new forest plantations and forest restoration. However, procedures have yet to be negotiated. To the extent that longer-term trading might reduce deforestation, especially tropical deforestation, it also will reduce loss of biodiversity and other local environmental services, such as watershed protection.

In the long run, the world's interest in abating greenhouse gas emissions is likely to remain strong, and forests can and should play an important role. Thus, it is imperative that the international community fund forestry activities that reduce net carbon emissions. This necessity suggests that sooner or later market-like mechanisms are likely to be created to encourage forest-based carbon storage or abatement. Hence comes the motivation for the World Bank to take a pioneering role in the field of "carbon forestry" and to engage continuously in developing and improving equitable and environmentally responsible forest carbon markets and adaptation measures in forestry.

Fostering Linkages between Poverty Reduction and Forest Conservation Strategy

The concerns with biodiversity conservation that have shaped so much of GEF, Bank, and World Bank/WWF Alliance assistance for forests have been primarily about global conservation values. They reflect predominantly Northern concepts and donor preoccupations and, in some situations, have constrained the pursuit of forest uses that can alleviate poverty. The Bank's revised strategy is based on the recognition that forest degradation and removal are not necessarily a consequence of poverty. They can be precipitated just as readily by rising incomes. In addition, the poor often conserve rather than overuse the resources that they manage. Therefore, approaches to SFM should not be based on assumptions of a mutually reinforcing "downward spiral" of poverty and forest degradation that can be halted only by limiting or preventing use of the forests by the poor.

Through ESW, policy dialogue, and its support to Protected Area programs, the Bank will promote improved understanding of the linkages among human activity, landscape change, and biodiversity preservation. Much of what might be considered by ecologists and foresters to be degradation or depletion of a forest resource can be considered by those depending on it for inputs in their livelihood systems to be transformation and even improvement of the resource.

There is, therefore, a need for greater appreciation among external stakeholders that the poor experience their own environmental problems, which need to be addressed separately from environmental policies seeking to satisfy concerns about global values. To address these local concerns, the Bank will seek opportunities to combine global macro initiatives with a more situation-specific focus. Such linked initiatives should reflect the protective mechanisms that local users themselves adopt and the attributes of a resource that they value and seek to conserve. As an example, recent Bank-supported initiatives in Costa Rica are moving in this direction.

Because a large part of the remaining tropical forest genetic resource exists in managed landscapes rather than in formally Protected Areas, it will be logical to focus more of the international community's conservation effort on sustainable management of what is in use. Many of these locally managed resources have high levels of biodiversity. However, in developing these arguments, it needs to be stressed that the Bank is not advocating downgrading the importance of biodiversity conservation. Rather, its concern is to better focus its approach on effective conservation. In recognition of the importance of forest uses for poverty reduction, conservation objectives for forests of value to local people appropriately could shift from a predominantly protective orientation toward encouraging sustainable systems of producing livelihood benefits in as "environmentally benign" a way as possible. This shift can be made, for example, by encouraging options that result in landscapes such as those found in parts of Southeast Asia that maintain a patchwork or mosaic of agricultural and agroforestry systems. While less species-rich than forests, such landscapes preserve much more biodiversity than the alternatives of plantations or clearance to crop agriculture. The Bank-supported Meso-American Biological Corridor is one example of this approach.

ALIGNING REGIONAL PROGRAMS WITH CORPORATE OBJECTIVES

The key challenge to improve the Bank's performance in forests along the lines described in this chapter is to link global priorities with the country initiatives of its Regional departments. The best approach to this task is improving coordination among Country Departments and Regional sector units, Regional operational units, and the groups in the Bank responsible for establishing and maintaining external partnership arrangements, special funding initiatives, safeguard policy application and monitoring, and knowledge management.

The Bank's operational units face severe budget constraints and will need to find cost-effective mechanisms to add these ingredients to their normal workloads. Choices will have to be made on how to focus budget and staff time on activities that can reap the highest benefits for poverty reduction and sustainable development. It is likely that incremental assistance to individual task managers through the Bank's system of Networks will be needed to achieve this emphasis.

This linked approach implies both a need for more focused and timely ESW and a strong preference for integrating forest issues and activities in larger cross-sectoral or economywide programs wherever possible. The operational Regions of the Bank have given considerable thought to these issues. As part of the Forest Strategy development process in the Bank, each Region assessed its previous programs and framed its own Forest Strategy for discussion in the Regional consultations. Following those consultations and the development of the three pillars of strategy and the financing proposal presented here, each Region prepared an outline of its current portfolio and reviewed the issues involved in building a program consistent with the three strategic pillars.

At this stage, the real prospects for increased investment in forests have yet to be assessed. In particular, willingness to borrow will be influenced by investments made in ESW (defined broadly to include consultative work) and the extent to which the Bank succeeds in the strategy of drawing donor grant-based assistance and focused investment from the private sector into the forest portfolio (chapter 4). Further work will need to be done with the Regions to expand their portfolios, based on country demand among other factors.

Africa

More than 70 percent of the population of Sub-Saharan Africa (SSA) is rural and depends on forests and woodlands for its livelihood. As much as one-fifth of the daily livelihood needs for rural families comes directly or indirectly from forests, including 20 percent of the disposable income used by the landless and poor families to pay for school fees and meet other family needs. Woodlands and forests supply approximately 60 percent of all energy (including industrial) used in Africa, as well as building materials and domestic energy for about 80 percent of all Africans and the totality of the rural population.

At various times throughout the past decade, forest-related activities have accounted for at least 10 percent of the GDP of 19 (forest-rich and forest-poor) African nations and for more than 10 percent of the national trade of 10 nations. Because of poor governance, the share of economic benefits remains highly skewed in favor of vested-interest groups, often depriving the state and the rural people of their fair share.

From a global biodiversity perspective, Africa is home to 25 percent of the world's remaining rainforests and contains 20 percent of the world's biodiversity hot spots. Four African countries feature among the world's 17 "mega-biodiversity" countries. The Central African forests alone store more than 23 billion tons of carbon, making them a critical buffer against global climate change.

Contrary to overall growth trends in the Bank's portfolio for Africa as a whole, the Africa Region's forest-related portfolio has been declining for more than 15 years. Current operations are of good quality, but the Bank's overall intervention is considered too small to have a meaningful effect.

Bank support to the Africa Region is shifting from traditional projects to programmatic operations that emphasize policies, institutional capacity, and partnerships and that are based on broad consultation, country ownership, and national and local political understanding and support.

Poverty reduction. A strategy will focus on integrating forests and biodiversity-related dimensions in broader Bank-supported rural-development policies and lending. In forest-poor countries, interventions will emphasize tree management and tree planting as part of policy and technology packages designed to help small farmers and generate income for the landless poor in the management of woodlands. In forest-rich countries, the emphasis will be on protecting rights (traditional and modern) and continued access to forest resources by those who depend on forests for subsistence, especially during the lean season. Emphasis also will be placed on creating frameworks enabling the poor to become partners with external investors in sustainable timber harvesting and to establish local hunters/wildlife management associations to use wildlife and other nontimber forest products sustainably.

Progress will be measured by the number and performance of new operations and programs integrating forests in broader Bank activities (especially community-driven development, natural resources management) and social monitoring of the poor's access to forests and of poor communities' participation in forest management and use.

Integrating forests in sustainable economic development. Forests will be integrated in sustainable economic development primarily through interventions in governance. These include improving core functions of government forest institutions, privatization, integration of forests in decentralization processes, fighting corruption, and empowering communities. To ensure that broad-based, self-sustaining support coalesces around the environmental governance agenda, politically difficult issues such as concessions and taxation reforms, tree tenure, and improved revenue collection will be tackled using public information, independent observers, participation, and partnerships.

Progress will be measured by successfully implementing institutional and policy reforms and improvements, increasing employment opportunities and fiscal forest revenues to the state and local communities, increasing transparency, and reducing corruption.

Protecting global and local forest values. Bank interventions will help reconcile long-term global environmental goals with short-term national economic needs and local priorities. While promoting market-based instruments to demonstrate that environmental services, certified products, and processes indeed can accrue economically important benefits, the Bank also will create opportunities for direct international resource transfers dedicated to conservation. Global initiatives such as transboundary parks and Protected Areas will be used to spearhead subregional integration through the harmonization of environmental policies and implementation of environment-related conventions (biodiversity, desertification, climate change).

Progress will be measured by the flow of funds into successful protection activities and the emergence of successful ecosystem market development programs.

East Asia and Pacific

The Bank's forest program is approximately 7 percent of total lending in this region: in relative terms, significantly larger than in other regions. In most cases, a sectorwide approach is being taken, strongly integrated with rural development and natural resources management programs. A major Bank concern in the natural forest component of its program has been the prevalence of the concession system of management, which has produced great pressure on the resource and social and environmental problems in many places. Governance issues are at the core of this problem. In addition, the rapid trend toward decentralization in the countries of the region will have consequences on managing the complicated, interconnected natural resource systems that will not necessarily be positive.

Poverty reduction. The focus is on the role of forests in the broader rural economy, community development, and user groups' participation in forest management. Resource expansion has been and will remain a major activity and will continue to advance toward management by communities and small farmers.

Progress will be assessed by the presence of forest issue activities in larger cross-sectoral programs, community ownership, and management of new resource projects.

Integrating forests in economic development. Major activities are being undertaken to address governance. These activities are focused principally on illegal logging operations, primarily in Cambodia and Indonesia. Policy analysis and reform are being pursued, especially in China, given its rapid change in regulatory and institutional structures. Land tenure reform is a particularly important issue in natural resources management in general throughout this region. The application of Bank instruments such as guarantee operations could be considered. In addition, the incorporation of forests and natural resource issues in structural adjustment and other broad economic programs and operations have been and are being implemented.

Progress will be measured by developments that help address the illegal log trade, introduction of policy and regulatory reforms, changes in forest land tenure, and treatment of forest issues in broad economic interventions.

Protecting global and local values. The focus will be to develop additional grant-based financing, work more effectively with partners, and apply the safeguards to achieve protection results.

Progress will be measured by the development of financing for Protected Areas and safeguard monitoring.

Europe and Central Asia

Primary issues of concern in forests in this region are the weakening of forest institutions and industries through rapid political change and conflicts; the impacts of changes in resource ownership, especially the restitution issue in some countries; and major needs for upgrading, training, and education throughout the region.

Poverty reduction. The primary focus will be recovery of forests and forest industries that have fallen into major disrepair and have contributed to a major intensification of poverty, and more effective engagement of civil society in collaborative forest management.

Progress will be measured by success in the revitalization of forest activities and industries and social assessment of the poor's engagement in forest management and protection.

Integrating forests in economic development. Close attention will be given to the impacts of macroeconomic reforms and adjustment, including measures taken under European Union (EU) accession; transparency and governance; improved investment climate for environmentally responsible and socially beneficial forest industries; major privatization issues in the sector; and improved forest management and protection, including investments in fire and watershed management.

Progress will be assessed by monitoring inclusion of major forest issues in adjustment, PRSPs, and CAS preparation, as well as in governance and public sector reform programs. Forest industry development and forest management outcomes will be tracked.

Protecting global and local values. The focus will be forest-poor areas and direct contributions for protection and conservation activities, especially those involving the poor. In forest-rich regions, the focus will be to develop effective markets for ecosystem services to generate greater national and local benefits.

Progress will be monitored in terms of incremental GEF and other grant-based financing for protection and conservation activities and evidence of developing markets for ecosystem services.

Latin America and the Caribbean

In recent years, the forest portfolio in this region has declined; yet, significant potential exists to invest in forests. The Latin America and the Caribbean region (LAC) contains the largest remaining areas of tropical moist rain forest, and the dependence of local and poor communities on forests of all types is high.

A major joint paper by the World Bank and Imazon in Brazil indicates that, in 83 percent of the Amazon forest, using land for agriculture yields low returns and has limited potential, so that, from the national economic point of view, sustainable management of forests is a much preferable use of the Amazon (Schneider and others 2002).

Poverty reduction. Some 40 million indigenous people—the majority of them poor—are concentrated in forest areas in the LAC region. The Bank's strategy is, with governments, to develop the policy, institutional, and legal bases to protect indigenous peoples' rights and access to forests; empower poor and marginalized groups; develop tenure security; and integrate agroforestry and secondary forest restoration activities involving the poor in rural development programs.

Progress will be measured by implementation of policy and legal reforms and social assessment of outcomes and by increases in engagement of the poor in forest aspects of larger cross-sectoral programs.

Integrating forests in economic development. Increased emphasis will be given to integration of forest management in major cross-sectoral programs; governance and illegal operations issues; development of awareness of the link between broad economic reforms in trade and fiscal areas to forest outcomes; and catalytic investments in SFM.

Progress will be measured by the presence of forest issues and activities in cross-sectoral programs, developments in governance and illegal removals, and field investments in SFM.

Protecting global and local values. The focus will be Bank involvement in GEF and other grant-based support to Protected Areas, the building of markets for ecosystem services, and evaluation of cross-sectoral impacts on forest areas.

Progress will be assessed by the initiation of functioning Protected Areas, evidence of ecosystem service market development, and ESW on cross-sectoral impacts.

Middle East and North Africa

Forests in the Middle East and North Africa region are scarce but vital for watershed protection and landscape outcomes. For this reason, forests need to be thoroughly integrated in natural resources management and sustainable rural development as a whole.

Poverty reduction. The priorities are to increase employment of the poor in reforestation, increase conservation and rehabilitation activities, build stakeholder incentives to manage resources sustainably, and develop local participation in these processes.

Progress will be measured by social assessments of the poor's participation in forest activities.

Integrating forests in economic development. The priorities are to incorporate forest issues in comprehensive development frameworks; develop linkages to forests in cross-sectoral investments, especially in rural development and water management; and undertake additional ESW on the linkages between forest land management and broader economic, rural, watershed and grazing policy, and institutional reforms.

Progress will be measured by the presence of forest issues and activities in large development programs.

South Asia

Increasingly, the focus in South Asia is better use of major inputs for rural development (especially land and credit) and development of better markets (trade and agroprocessing issues), as well as public sector institution reform (agriculture, water, forests, energy, health, finance, and infrastructure), particularly to achieve better service delivery to the poor. The forest program is closely tied to this broad approach to rural development and has the following objectives: (a) to increase benefits to the rural poor through empowerment, ensuring that they control natural resources through rights to ownership, access, management, or usufruct and maximizing the value of the resources through efficient markets; and (b) to provide environmental services for all sections of society through a managed viable system of sustainable use, increased productivity, and improved conservation of natural resources. The forest program will operate through rural development and natural resources management programs on issues such as empowerment of the poor, SFM, and improved local community control over natural resources use, rights, and markets.

Poverty reduction. Building on a long history of engagement in community forestry in South Asia, the Bank will continue to emphasize creating an enabling environment for poverty reduction, improving community participation and institutions, increasing returns to forest activities (through productivity and joint forest management), and reducing vulnerability by securing access and diversifying ownership of the resource asset base toward the poor.

Progress will be measured by continued social and implementation assessment of joint forest management activities.

Integrating forests in economic development. The focus will be on improving governance (especially, correcting major distortions in incentives and markets that reduce the value of the forest resource and experimenting how to adequately pay communities for providing ecosystem services), reforming the public sector (including responsiveness, accountability, and fiscal soundness of institutions), and developing efficient markets and competitive private sector producers in the sector.

Progress will be assessed by the policy and institutional reforms designed and adopted, performance of local forest product and ecosystem service markets, and changes in relative private sector and public sector roles in forestry-related activities.

Protecting global and local values. South Asian governments have emphasized the interests of poor rural communities in international fora, such as those on the climate change, biodiversity, and desertification conventions. The regional strategies' emphasis on poverty reduction and sustainable livelihoods provides further backing to this stance and supports meaningful action in the field. These actions may include the conservation of biodiversity in mixed-use forest settings, adding rigor to fiscal and poverty impact analysis, strengthening participatory performance monitoring, and more effectively addressing the impacts of nonforest activities on forest areas.

Progress will be assessed by the further development of mixed-use forest management systems that successfully address local livelihood as well as global biodiversity and global warming concerns, and as well as the development of cost-effective assessment and monitoring systems that meet adaptive management requirements, document local and global impacts, and more effectively address land-use management and social and environmental assessment needs.

CHAPTER THREE

Implementing the Strategy: Building Effective Partnerships

ISSUES IN IMPLEMENTATION

The new Forest Strategy proposed here will require special emphasis on implementation. In its review of the 1991 Forest Strategy, OED not only was critical of its content but also was particularly concerned about the failure in implementation. OED recommended that, to fulfill both its global and country roles, the Bank be proactive in establishing partnerships with all relevant stakeholders; that it align its resources with its objectives in the forest sector; and that it use its "global reach" to address mechanisms for, and mobilization of, international concessional resources outside its lending activities (see appendix 1 for a specific response to OED's recommendations and appendix 6 for the full text of those recommendations).

Implementation of the proposed strategy will emphasize selectivity in areas of focus so that resource use will more closely match objectives and partnerships that generate effectiveness both internationally and in client countries and mobilize financing both externally and internally. In addition, the content of the strategy is built around Bank strengths: analytical capacity and ability to convene multiple stakeholders. Although implementation will remain a major challenge requiring innovation and facing risks, the rewards in reducing poverty and more effectively preserving environmental services will be significant.

Building Effectiveness

Building the Bank's effectiveness to achieve the objectives discussed in this document will require

- Addressing the global challenges and realities for forests and the Bank's role; then refocusing the Bank on engagement in the sector on the basis of the priorities embodied in the three pillars of the revised strategy (chapters 1 and 2);
- Addressing the major cross-sectoral and macroeconomic linkages and sector policy issues (chapter 2);
- Developing the necessary partnerships with donors and other stakeholders to leverage coordinated efforts being made in forests;
- Attending to the internal Bank institutional issues and incentives so that these will (within a prudent and realistic framework) encourage rather than work against the overall directions of the proposed Forest Strategy.

This chapter will discuss the development of the partnerships that will be needed to maximize the Bank's effectiveness in forests, and it will introduce important internal institutional issues. The internal incentive implications related to these topics will be discussed in chapter 4.

DEVELOPING THE PARTNERSHIPS NECESSARY TO IMPLEMENT THE STRATEGY

Donor and National Stakeholder Partnerships

Past failures by the donor community to effectively address deforestation and related issues are partly a reflection of the fragmentation of individual donor priorities. They also are related to the reluctance of some governments and donors to engage in policy dialogue on politically sensitive issues such as the negative social and environmental consequences of powerful vested interests. Therefore, this strategy should give major emphasis to working with national and local partners on strong, consensus-based agendas. When vested interests and very large financial gains are at stake, the potential to leverage solutions depends on the agreement of all the *major* stakeholders on a shared strategy for forests. Leveraging solutions also depends on developing the necessary consultation and dialogue with actors at all levels to ensure that the agreed strategy can and will be implemented. Donors in particular will need to agree on the broad objectives of their engagement in forests. Even more importantly, donors will need to agree on the specific steps and investments to be made through a carefully developed engagement strategy. To succeed, this dialogue and approach will require the strong support and ownership of governments and other stakeholders. Building this dialogue and consensus will be one of the major challenges of the partnerships.

At the global level, the Bank should continue to participate in fora that enable it to work in partnership with other organizations of global reach and within the framework of international agendas. The UNFF should provide a useful core framework. The UNFF was created under the Economic and Social Council of the United Nations in October 2000, after a 10-year process of intensive discussions among governments. UNFF's objectives are to "promote the management, conservation, and sustainable development of all types of forests and to strengthen long-term political commitment to this end" (box 3.1). UNFF offers an intergovernmental platform for policy dialogue through which difficult issues, such as good governance in the sector, can be reinforced by a broad intergovernmental commitment and effective monitoring and progress assessment. It offers an intergovernmental forum in which countries can put forward their pledges and accomplishments in SFM. It is a forum through which the Bank can support the international commitments of its client countries to promote the implementation of IPF/IFF Proposals for Actions while working closely with other institutions, particularly in the areas in which the Bank has a comparative advantage.

In parallel, and supportive of the UNFF, is the newly formed CPF, which will contribute substantively to the work program of the UNFF. CPF consists of major multilateral organizations such as the World Bank, other UN agencies, research organizations, and other forest-related organizations.[10] CPF's role will encompass facilitating and promoting coordinated and cooperative action, including joint

BOX 3.1

Functions of the United Nations Forum on Forests

The ECOSOC resolution assigns the following functions to the UNFF:

- Facilitate and promote the implementation of the Intergovernmental Panel on Forests/ Intergovernmental Forum on Forests Proposals for Action as well as other actions that may be agreed.
- Provide a forum for continued policy development and dialogue.
- Enhance cooperation as well as policy and program coordination on forest-related issues.
- Foster international cooperation.
- Monitor and assess progress at the national, regional, and global levels.
- Strengthen political commitment for the management, conservation, and sustainable development of all types of forests.

Source: United Nations 2000.

programming and submissions of coordinated proposals, and facilitating donor coordination. The CPF will submit coordinated inputs and progress reports to the UNFF and undertake periodic reviews of its effectiveness. The World Bank will continue to take reporting responsibility in the specific fields in which it has recognized comparative advantages.

The Bank already is operating within comprehensive development frameworks that establish the overall development and policy agenda for a country.

The concept of NFPs with governments and other stakeholders has been developed through recent major intergovernmental dialogues on forests: the IPF and the follow-up IFF, which were the forerunners of the UNFF. These processes have the potential to develop strong political commitment to reform and development of the forest sector. In the field, an NFP is convened by a government, supported by a group of donors, and based on consultation with a broad group of stakeholders. The purpose of an NFP is to develop a consensus on how forests are to be used, by whom, and for what; and then to implement that agreement.

The Bank could become involved in this process in selected countries in which the potential exists to develop and refine programs that are highly compatible with the Bank's CDF. The Bank could provide an organizing mechanism for governments, donors, NGOs, and the private sector to work together on sector matters and the linkages to broader economic and environmental issues. This collaborative approach will amount to an *enhanced* NFP. Where it is possible to achieve, it could be a productive mechanism to align, select, and prioritize donor activities in the forest sector. The collaboration could have a longer-term objective of improving the performance of donor programs in forests, thus enhancing the willingness of donors to raise their future contributions to the sector. However, it must be emphasized that, at this stage, the Bank probably could use this approach only on a highly selective basis.

Two major aspects of developing and enhancing NFPs are a challenge for the Bank and its partners: (a) integrating cross-sectoral approaches in NFP formulation and implementation processes; and (b) coordinating Bank programs and priorities with broader donor programs. NFPs will need to be comprehensive, covering forest issues ranging from cross-sectoral policies and investments in forest management to establishing and sustaining Protected Areas. The implementation of these NFPs will require blended financing from multiple sources ranging from funds at commercial rates to funds at highly concessional terms.

BOX 3.2

Principles of National Forest Programs

Principles of NFPs were determined at the fourth session of the IPF in 1997. They include

- National sovereignty and country leadership;
- Consistency with national policies and international commitments;
- Integration with the country's sustainable development strategies;
- Partnership and participation;
- Holistic and intersectoral approaches.

In addition, at the same session, the IPF adopted the following specific elements to be considered during the development and implementation of NFPs:

- Appropriate participatory mechanisms that should involve all interested parties
- Decentralization, where applicable
- Empowerment of regional and local government structures consistent with the constitutional and legal frameworks of each country
- Recognition and respect for customary and traditional rights of, among others, indigenous peoples, local communities, forest dwellers, and forest owners
- Secure land tenure arrangements
- Establishment of effective coordination mechanisms and conflict resolution schemes

Source: United Nations 1997.

The PROFOR Initiative. The analytical work that will underlie NFPs is critical to the implementation of the proposed strategy. Since knowledge is an important public good, it requires being developed in partnership and shared among stakeholders. One important means by which the Bank could pursue this objective would be to host a multidonor partnership on forests.

An existing donor forests partnership—PROFOR—which has been moved from its previous location at UNDP to the Bank, is a possible mechanism.[11] Bringing the PROFOR arrangement closer to the Bank enhances the focus on three aspects of NFPs that closely match the Bank's interests: cross-sectoral policies, governance, and financial instruments. The PROFOR donors strongly favor moving it to the Bank because this would enable the PROFOR analytical resources to combine with the Bank's strengths in analytical and sector work and country dialogue. PROFOR will supplement the World Bank's operations in ESW and support governments in their policy reforms and planning. Specific provision will be made for Bank Regional operational staff to work with PROFOR both on the definition of priority countries and tasks and then on implementation of the work program. The terms of reference and objectives for PROFOR would ensure that each supporter obtains value from the arrangement. To ensure key stakeholder representation, an oversight committee would be established to guide activities and monitor progress. The Bank will need to provide an effective linkage between the PROFOR program and the Bank's sector portfolio and to manage ESW, identification, and preparation activities so that major improvements in coordination result from the process.

Activities sponsored by the PROFOR Initiative will be based on the three objectives listed in chapter 2: (1) harnessing the potential of forests to reduce poverty, (2) integrating forests in sustainable economic development, and (3) protecting vital local and global forest environmental services and values. The PROFOR team will work closely with Bank operational staff, client governments, and other stakeholders to select the specific analytical and knowledge activities to be carried out. The team would ensure that this analytical work is integrated and coordinated with broader cross-sectoral and economywide interventions. Opportunities would be sought to incorporate the results of enhanced sector analysis and consultation, and the issues they raise, in Bank CASs, business plans, and forthcoming PRSPs; adjustment operations; and major cross-sectoral programs in rural development, natural resources management, and infrastructure.

Total donor support for PROFOR is expected to be in the range of US$13 million to US$15 million over five years. This amount would permit a dedicated group of experts to support Regional staff in ESW and the development of sector programs through NFPs. In their support to improved NFP processes, the PROFOR team would focus on four critical thematic areas:

- Building institutional, legal, and policy structures to enhance the contribution of forests to sustainable livelihoods.
- Building and reinforcing the relationship between improved governance and forest sector development.
- Assembling the institutional and legal framework for national financing strategies for sustainable forest sector development, including analyzing innovative financing mechanisms for all the goods and services that can be produced from well-managed forests.
- Analyzing the cross-sectoral and macroeconomic effects of policy interventions and investments on forests.

Other international arrangements. As outlined earlier, successful implementation of enhanced NFPs will require the active participation of all stakeholders through a consultative process. Consultation should be ongoing throughout implementation. It will be necessary to bring together national and international research and technical groups. The main international collaborative research organizations will be CIFOR, ICRAF, the European Forest Institute, FAO, and ITTO. In addition, research and technical support will be sought from international NGOs such as IUCN, WWF, World Resources Institute (WRI), and International Institute for Environment and Development (IIED).

Collaboration with FAO. FAO is establishing an Implementation Facility to provide technical support to strengthen in-country capacity for developmental work on NFPs. It will be important for the Bank Group and FAO to work together and with other major partners on this initiative, using their respective comparative advantages. For the past 30 years, the Bank has had a strong collaborative partnership with FAO, especially through the partially Bank-financed FAO/IBRD Cooperative Program, which has played a key role in supporting Bank in-country agriculture and forest-related sector studies and project identification/preparation activities. The PROFOR Initiative would provide an important vehicle in which the two agencies could collaborate. For the focus group of countries, working closely with the government and other stakeholders, PROFOR would use the resources and comparative advantages of the partners to formulate a consultative and analytical process in each country. This process then would form the basis of an NFP. The government would be in a position to identify issues and priorities and a comprehensive request for an NFP to support the broad-based program of reform and development.

CGIAR. Through its support to the CGIAR system, the Bank has developed strong partnerships with the Indonesia-based CIFOR and Kenya-based ICRAF. The emerging research results of these and other CGIAR centers will be integrated in Bank-supported in-country sector work.

Other research institutions. The Bank has a global role in sharing knowledge and supporting research on forest issues. While the Bank has limited comparative advantage in forestry and the technical aspects of forest management, it can be a vehicle to produce and share knowledge in the areas of cross-sectoral impacts as well as economic and policy analysis. The Global Forest Information Service, coordinated by the International Union of Forest Research Organizations (IUFRO) and FAO, has the potential to support these activities. Although not supported directly by the Bank, the work of several other specialized technical and/or policy research institutions has enhanced Bank staff understanding of forest-related issues, constraints to progress, and potential solutions. These institutions include, for example, IUFRO, the European Forest Institute, the Iwokrama International Centre for Rain Forest Conservation and Development, WRI, and IIED. Periodically, the Bank also has drawn on its linkages with universities that are pursuing research on topics of mutual interest.

GEF. The GEF brings together more than 165 member governments, development institutions, and the scientific community as well as a wide range of nongovernmental and private sector organizations to work toward the common goal of global environmental protection. It facilitates technical and information linkages between GEF's focal points and projects. The GEF has been a close partner of the Bank in helping to develop the Bank's forest and biodiversity portfolio, which amounts to about US$380 million in more than 50 countries.

ITTO. An active dialogue is ongoing about the possibility for closer collaboration between the Bank and the ITTO. This intergovernmental organization is itself undergoing reform. It has adopted a series of targets that fit well with some aspects of the Bank's own programs, including (a) a 15-million-ha Protected Area target by the year 2003 for transboundary protection areas in the tropics; and (b) a forest management target of bringing 500 forest enterprises under ITTO's Criteria and Indicators scheme by the year 2003. This second target includes the establishment of an independent monitoring system. A third target is a total of 10 million ha of restored forest ecosystems by the year 2003.

Nongovernmental Organization Partnerships

World Bank/WWF Alliance. In 1998 the Bank joined forces with WWF to form the World Bank/WWF Alliance to promote both the conservation of forests and SFM. This partnership has embraced the World Bank's goal of ensuring sustainable livelihoods for forest-dependent poor and indigenous peoples and was founded to achieve three global targets by 2005 (box 3.3). The Alliance will be an important partnership in implementing the Bank's Forest Strategy and Policy. The Alliance's ambitious targets challenge not only the Bank and WWF, but also other actors—governments, business, conservation groups, local communities, and aid agencies—to take action to reduce the loss and degradation of forest resources.

The partnership strives to achieve its goals and objectives through strengthening policies and institutions and bringing together stakeholders to facilitate their work: building capacity for improved management and providing the framework for independent certification of forest management both in market and nonmarket contexts.

The Alliance is moving into implementation of its Protected Areas and forest management targets. It is doing this through pursuing:

- Activities in key forest countries in which the Alliance works with governments, NGOs, and the

BOX 3.3

World Bank/WWF Alliance Targets for Forest Conservation and Improved Management

Targets for 2005:

- 50 million ha of new forest Protected Areas.
- A comparable area of existing but highly threatened forest Protected Areas secured under effective management.
- A global target of 200 million ha of production forests under independently certified sustainable management.

The initial global targets for forest conservation and improved forest management were designed to be significant considering the scale of the problem but also had to be achievable within a reasonable time frame (seven years).

Conservation targets. Of the 3,300 million ha of forests remaining on Earth, only 6 percent are in legally Protected Areas. Many of these legally Protected Areas exist only on paper, without effective protections on the ground. Hence, the targets would increase the global current area under IUCN protection categories I–III by more than 25 percent.

Improved forest management targets. These targets aim at improving the overall biological, social, cultural, and long-term equity and economic outcomes of existing forest utilization operations. Targets would be achieved through a range of approaches from large-scale landscape analysis to local projects. Achieving these targets would mean demonstrably improved management in approximately 30 percent of production forests worldwide.

Source: World Bank/WWF Alliance 1999b.

private sector to develop programs and projects that can bring about commitment and change at the scale required by the global targets.

- Activities designed to enhance integration of the Alliance targets and activities with "upstream" investments. These investments are aimed at generating future interventions and resources to improve forest management and facilitating key stakeholder groups within countries. These groups are expected to develop and apply standards for forest management certification and assess and improve the effectiveness of Protected Area design and management.

Through the Alliance, WWF and the World Bank are committed to finding ways to achieve more than either organization could accomplish independently. By combining the World Bank's financing operations, access to policy dialogue, convening power, private sector partnerships, and analytical capacity with WWF's knowledge of conservation science, local field presence, large technical skill base, public trust, and market influence, the Alliance is a mechanism formed to tackle forest management issues much more broadly. It will require additional support and partners to maximize its ability to draw on the strengths of each institution and to mainstream dialogue on important issues, including independent monitoring and certification of forest operations.

Other NGO partnerships. Other NGO partnerships also are being developed and will need to be strengthened.

Forest Trends is a Washington-based NGO created in 1998 with support of the World Bank and the MacArthur Foundation. Forest Trends grew out of the earlier multistakeholder, Bank-supported Forest Market Transformation Initiative. Forest Trends has brought together a potentially influential coalition through its board of directors of Northern and Southern private sector entities and NGOs actively engaged in the forest sector.

Forest Trends focuses its work in two strategic directions. One is to develop markets for forest ecosystem goods and services and SFM that will lead to healthy and expanding forest ecosystems and sustainable livelihoods for forest-dependent peoples. The other is to improve the efficiency of forest product use and the development of alternative sustainable sources of fiber to take the pressure off natural forests. Forest Trends has increased private sector understanding of forest carbon market opportunities and the challenges of creating a market in *forest carbon offsets* to support forest conservation and sustainable use. Forest Trends also has convened industry leaders, analysts, and advocates to evaluate and promote dialogue and investment in nontimber forest goods and services.

Critical Ecosystem Partnership Fund (CEPF). The CEPF is a major new source of international funding directed primarily to nongovernmental, community, and grassroots organizations. Within 21 "hotspot" ecoregions, it seeks to engage all stakeholders that can best contribute to solutions for biodiversity conservation. The CEPF is a partnership among Conservation International (CI), the GEF, the MacArthur Foundation, and the World Bank.

Closer collaboration between the Bank and the IUCN is being developed. The IUCN has been a major partner with the Bank in developing the revised Forest Strategy and Policy. Closer cooperation with IUCN could greatly strengthen the Bank's approaches to collaborative forest management, achievement of SFM and Protected Area targets, and strategies to more effectively contain forest fires. The revisions proposed in this strategy for a more integrated approach to forest conservation and development are highly compatible with the priorities outlined in the Forest for Life Strategy jointly developed by IUCN and WWF.

Possibilities also are being explored for developing collaboration with The Nature Conservancy (TNC), a leading conservation agency, which has been collaborating closely with the Bank in developing innovative financing approaches for forest conservation.

Collaborative possibilities also are being explored with a WRI-sponsored initiative, Global Forest Watch, which is an independent network of national and local organizations engaged in monitoring and mapping logging, mining, road building, and other developments that impact forests in major forested regions of the world.

Improving Links to the Private Sector

Participation of the international private sector is especially important to produce favorable forest outcomes. International flows from private interests to forest operations and forest-using industries dwarf the combined flows from multilateral development banks, donor agencies, and NGOs. There are major opportunities to leverage more private sector investment into sustainable forest activities, or related investments (for example, in processing), especially in situations in which the environment for responsible investment is not favorable.

CEOs Forum on Forestry. In 1998, the president of the World Bank initiated the CEOs Forum on Forestry, a dialogue originally initiated with the chief executive officers (CEOs) of large forestry firms from around the world. This arrangement has since expanded to include retailing and production firms and some major international and Southern NGOs. The forum has become an important vehicle through which the Bank and others can exchange views and develop ideas on how to approach forest issues, to partner more effectively with the private sector in its programs, and to more actively incorporate the private sector in Forest Policy dialogue with national and international stakeholders. As a direct result of the CEOs Forum, working groups on specific issues and regions have been initiated. The CEOs Forum Africa Working Group has become a major initiative in the Congo Basin, bringing together major international forest companies, international and local NGOs, representatives from science, and the African Timber Organization. The group focuses its work on codes of conduct, bush meat poaching, reduced-impact logging, biodiversity conservation in forest concession areas, and issues in sector governance and timber trade. The Forest Dialogue Group, another spin-off of the CEOs Forum, seeks to maintain a global dialogue on important issues in forest development among industry, NGOs, forest owners, and forest workers associations.

Need for additional partnerships. A new partnership with the private sector, civil society, and governments needs to be established for forest carbon. All forests store carbon, and this carbon is released when they are disturbed or burned. Up to one-third of greenhouse gases being emitted into the atmosphere are estimated to come from biomass burning, clearing, and related activities. Protection of threatened forests and reduction of the impacts from logging and agricultural development are potentially important ways of increasing the contributions of forest lands to carbon emissions. In addition, the creation or enhancement of new carbon sinks through tree planting has been suggested to offset some greenhouse gas emissions.

The clear potential to reduce emissions from forest lands and to use forest lands to sequester carbon stored in the atmosphere has led to considerable interest from some countries and companies in sponsoring forest conservation and management projects to offset their carbon emissions. However, while these offset projects could provide useful funds for forest conservation and management, they remain controversial. Many environmentalists fear that offset projects might encourage bad forestry practices, for example, by providing incentives to replace natural forests with fast-growing exotic species. It also is clear that forests alone can never be more than a relatively small but still significant part of an overall solution to the greenhouse gas problem and that early actions to reduce the fossil fuel emissions from industrialized nations at their sources remain essential.

The enhancement and improved management of "carbon sinks," including afforestation, reforestation, and forest conservation, are eligible activities under the Joint Implementation provision of the Kyoto Protocol. At the resumed Sixth Conference of the Parties in Bonn, July 2001, Parties to the UNFCCC agreed that afforestation and reforestation activities also were eligible activities for the achievement of emissions reductions under the Kyoto Protocol. The Parties decided that applicable rules would be agreed at the Ninth Conference of the Parties to the UNFCCC in 2004. The volume of certified emissions reductions achieved through afforestation and reforestation to meet UNFCCC obligations of the industrialized countries in the first commitment period of 2008–2012 is five times 1 percent of their individual emissions reductions obligations under the Protocol. The door is now formally opened to using the carbon-offset trade to stimulate reforestation and afforestation activities, but not as an incentive to avoid deforestation. Nevertheless, applied sensitively to environmental and social concerns, these eligible forest asset creation activities under the Kyoto Protocol can support improved forest management in developing countries.[12] Through public–private partnerships and with its experience in implementing the Prototype Carbon Fund (PCF), the Bank will contribute to the Parties' review of the rules governing afforestation and reforestation activities in the CDM.

The Bank's PCF is able to invest up to 10 percent of its subscribed capital—about US$15 million—in Kyoto Protocol–eligible carbon sequestration activities in transitional economies and developing countries. These PCF carbon purchase activities will enable the Bank to pioneer environmentally, economically, and socially responsible investment in forest conservation, management, and restoration activities in developing and transitional economies catalyzed by largely private sector finance. PCF resources are still modest in terms of the scale of required demonstration activities. Therefore, a "prototype forest carbon" or "sinks" fund embracing partnership with NGOs, industry, and governments seems warranted to extend "learning by doing" in creation of Protocol-eligible carbon assets from forestry in an environmentally and socially responsible way across a representative range of potentially reforested landscapes and contexts. However, the Bank's role should be catalytic only. Once sound performance benchmarks are in place and these markets have attracted private sector participation, the Bank's role will have been completed.

More generally, the Bank needs to consider the expanded opportunities for interaction with the private sector that are presented by increased private sector interest in investment in forests in Bank client countries. For example, a number of very large institutional investors are becoming interested in supporting responsible forest investments both in the large remaining natural forest areas of the world and in plantation activities. Institutional investors cite the findings of studies indicating that returns to forest investments are countercyclical to equities markets movements and therefore a useful hedge. They recognize the option values of forests, namely, that contrary to virtually all other commodities traded internationally, the real price of forest output has a long-term tendency to rise. There also are strong possibilities that the potential markets for carbon sequestration and other environmental goods and services from these forests will grow.

Because of the institutional origin of the funds they invest, the interest of these firms is firmly in the area of sustainable investments and social responsibility. Many governments have a strong interest in such investors entering their forest sectors. These investors bring management strength and access to reliable international funds, and, ultimately, they affect the credit standing of the countries in which they invest. What often is lacking in some countries with large forest potential is reasonable political risk security and the means to address problems or constraints with the investment-enabling environment. The Bank could have a major role in developing this potential, largely through its capacity to address issues of the investment-enabling environment in forest sectors. That is, the Bank could address some of the risk factors that presently inhibit these desirable forms of development.

The Bank Group and its partners are in a good position to address some of these difficulties because they have

- Access to government economic decisionmakers on the basis of long association under existing lending and other programs;
- Access to the guarantee instrument, which could be employed in creative ways to combine reducing the exposure of potential investors to policy risk with private underwriting of commercial risk;
- The means to incorporate specific policy reforms and undertakings needed to attract responsible private investors to the forest sector into larger economic restructuring and reform programs, such as those being developed under structural adjustment, poverty reduction, or other broad economic reform interventions;
- Field presence to include specific forest activities in larger rural development or natural resources management programs that could gain the attention and support of important local stakeholders and political figures resulting in positive forest outcomes.

Clarifying and supporting a framework to attract appropriate private sector investment in the forest and forest industry sector will facilitate IFC's role in financing new investments and investors in sustainable forest operations. IFC has the capacity to finance a wide range of operations from large investors to medium-sized enterprises through loans or equity investments made directly or through financial intermediaries. IFC's investments in private sector, sustainably managed forestry and forest industry projects have demonstrated and will continue to demonstrate the financial viability of independently certified, sustainable private sector forestry operations.

The Bank Group could follow a threefold strategy to develop this potential. First, it would raise its own profile within the framework of NFPs. Second, it would develop a sharper focus on the interests of forest-dependent people, improved forest management, and effective protection of forest resources in critical forest regions. To achieve this, the Bank Group would need to invest in partnership-based consultative processes. Third, it would emphasize bringing the message of the value of forests to all stakeholders.

There is a need for the Bank to deepen both the range of activities supported through its interaction with the private sector and its contacts in the private sector, especially to intensify the dialogue with some very large institutional investors that have expressed interest in working more closely with the Bank on investments in SFM in some key Bank client countries.

IMPROVING INTERNAL BANK WORKING RELATIONSHIPS AND ACCOUNTABILITIES

International Finance Corporation and Multilateral Investment Guarantee Agency

Two institutions of the World Bank Group, the IFC and the MIGA, offer additional important instruments for drawing the international private sector into investing in SFM in Bank client countries. Clearly, to implement this strategy, a closer working relationship between the IFC and IBRD/IDA needs to be sought. The ESSD Forests Team would have a role in bringing this about by building a closer relationship with forest-oriented staff in IFC. MIGA could have a larger role in forests through its ability to insure private investors against political and catastrophic risk.

World Bank Institute

The WBI has the capacity to bring awareness and training to policymakers in developing countries. By focusing on the thematic areas of forests and poverty, governance and economic management, and local and global markets, WBI can help lay the groundwork for the solid implementation of a new Forest Strategy and Policy. WBI also can help countries that wish to embark on Forest Policy reform programs. In countries in which programs and projects are being prepared, WBI can work in parallel with the Bank to support the intensive consultation and dialogue processes that will be needed to ensure that understanding and feedback on forest initiatives are broadened and that constituencies for change are developed.

Role of the ESSD Forests Team

The integration of forest-focused staff and resources from the Environment and Rural Development Departments of the ESSD Vice Presidency into a single forests anchor team in 1999 has provided a basis for coordinating policy and some global tasks for forests. The expanded activities implied for the Bank forest program in the new sector strategy suggest that the major priority tasks for this team are

- To be instrumental in improving coordination with IFC, MIGA, WBI, and GEF. However, to achieve this, the team must broaden its mandate beyond the IBRD/IDA forest community and develop specific strategies to engage with these groups
- To coordinate with PROFOR and assist in fulfilling its objectives, support the World Bank/WWF Alliance, and manage other major international partnership arrangements
- To take major responsibility, through the sector boards and by other means, for monitoring and advising Regional forest teams in the Bank on implementing the new strategy. To do this effectively, the team will need to develop partnership arrangements with the Quality Assurance Group
- To advise the Regions in applying and monitoring the new OP on forests

CHAPTER FOUR

Implementing the Strategy: Incentives, Selectivity, and Deliverables

The Bank's strategic approach to forests needs to be integrated in country programs through developing and costing interventions that can lead to meaningful levels of achievement consistent with country priorities. Such integration will need to be consistent with the Bank's organizational structure and decisionmaking. Special incentives may be necessary to ensure that the strategic objectives are achieved.

Decentralization of Bank decisionmaking, with its focus on demand-driven priorities established at the country level, has implications for the institution's engagement in the forest sector. On the one hand, decentralization is highly compatible with the emphasis in this strategy on integrating forest issues and activities in broader economywide and cross-sectoral programs and strategies and achieving country ownership. On the other hand, decentralization may strengthen the tendency already apparent in some countries to avoid proposing and borrowing for projects and programs with more risks and higher transaction costs, but whose importance to economic and environmental sustainability is substantial and whose benefits to the poor are high but possibly less immediate. This outcome is especially likely in cases in which forest projects also are perceived by Bank country teams and managers to involve intensive management of external relations. The average size of forest projects in the Bank is relatively small, but some project preparation costs are fixed; and safeguard compliance costs are higher in this sector than in many others. The net result is that overall transaction costs of engagement in forests are higher than the average for Bank budget allocations. These higher transaction costs and higher risks must be balanced against the importance of forests to reducing poverty and to the delivery of vital environmental services both locally and globally.

The Bank must find strategic ways to leverage its finance and use its influence to address global and country issues in the forest sector while recognizing the importance of country ownership and acknowledging the higher costs associated with managing projects and programs in important but sensitive sectors such as forestry.[13] The approach to implement the strategy proposed in this document is based on the following:

Adapting the new approach to global and corporate priorities being developed in the World Bank to enhance the commitment and resources devoted to the incremental PROFOR outlined in chapter 3.

Using incremental budget resources for ESW, thus enabling country teams to incorporate potential forest impacts more adequately into CASs, PRSPs, and cross-sectoral and economywide measures under consideration by the Bank. In the medium term, the funds will underwrite the (relatively higher) costs of preparing and supervising incremental investment in forests. ESW and analytical work will be front-loaded in the program to build the foundations for increased investments in the forest sector and to heighten awareness inside and outside of the Bank of forest issues.

Using this enhanced level of commitment by the Bank to leverage additional donor funds and stronger participation by other partners. These funds and participation will be channeled into better-aligned programs of assistance for the forest sectors in selected client countries. Enhanced funding and stronger participation by the Bank and other partners also will provide an opportunity to coordinate donor analytical work more effectively and to make better use of the comparative knowledge and analytical strengths of the major partners. The funds will involve support for some donor and NGO partnership initiatives that already are well advanced in conceptual design and the formation of other initiatives that are less well developed. Especially important will be strengthening relationships in client countries with compatible private sector investors (or potential investors) in forest management and protection and forest industry development.

Developing and maintaining an enhanced forest sector lending portfolio in the Bank. An increased ESW program is indispensable, to be leveraged by stronger interaction with borrower countries, donors, and other stakeholders with common interests in the forest sector (see section on Selectivity below).

BUILDING THE INTERNAL BANK COMMITMENT TO FORESTS

Raising the Bank's engagement in the forest sector will be initiated through incorporation of forest issues, where these are relevant and important, *well upstream* in PRSP and CAS discussions so that Country Departments can undertake the necessary follow-up analysis in a timely fashion. The basic objective of analytical work is to raise awareness of the potential and importance of dealing effectively with forest outcomes. As demand results from this process, additional resources could be committed to increase lending and nonlending activities in the sector.

Increasing Economic and Sector Work

The primary means to initiate the process of building analysis, awareness, and then demand for forest investments and for incorporation of forest issues in larger programs through the CAS and PRSP processes would begin with an enhanced commitment to focused ESW. As noted in chapter 1, ESW in the World Bank's forest program has declined precipitously. A major objective of the new strategy would be to rebuild ESW in forests and forest-related work to a level close to that which prevailed at the beginning of the 1990s, for a defined period, after which it is assumed that client demand will lead to higher allocations for this purpose that will be made autonomously through the normal country budgeting process. Under the broader goal of having forests take their place in an enhanced World Bank commitment to environmental commons, the two specific objectives of increased forest ESW are the following:

- To provide the knowledge base and analytical support to effectively integrate forests in development programs and to link policies, projects, and programs in other sectors to forest outcomes. Besides establishing the importance of forests in poverty reduction, sector work will determine the potential impacts on forests and forest-dependent peoples of forest and nonforest sector activities contemplated by the Bank through investment, adjustment, and broad poverty reduction programs. It will establish the alternative paths by which policy, institutions, and management of forest resources can be improved.
- To develop the analytical and consultative bases needed to build awareness and demand for forest investments. Sector work will provide the bases for effectively integrating forest programs in CASs and PRSPs not only by creating knowledge of forest outcomes but also by building government and public support for engagement. For sector work to build this awareness and demand, it must be participatory and widely discussed.

Using Partnerships to Leverage Impact

Donor partnerships to boost the ESW program. As discussed in chapter 3, hosting PROFOR will strengthen the Bank's international role in the forest sector. A primary spin-off of this multidonor partnership will be the addition of analytical outputs for country forest sectors that would well exceed the capability of any one of the partners to develop alone. Since the externalities of this sector work are high, it is expected that all donors, including the Bank, would contribute to the common effort of building knowledge through analysis.

Seeking Blended Financing to Address Local and Global Forest Issues

Forests are part of the environmental commons and, as demonstrated in chapter 2, offer many externalities to both local people and the wider national and international community. Besides addressing poverty and economic development, improving the quality of forest operations will generate benefits through securing future productivity, maintaining watershed quality, preserving biodiversity, and sequestering carbon in the environmental commons. Many client countries have argued that, because some of these key benefits are global benefits, the countries that provide them must be compensated through concessional funding mechanisms. In its review of the 1991 Forest Policy, OED discussed and agreed with this view.

In addition to the need for concessional resources, forest operations will need financing of sufficient magnitude to make a difference. Isolated forest projects will not always bring sufficient benefits to stakeholders to warrant their interest in conserving forest cover. Financing from other donors, the private sector, and NGOs will be needed to supplement government and Bank investments. This program of investment from various sources will need to be coordinated and have a unified strategic focus.

To leverage the impact of its own increased commitment to forests, the Bank will seek the active participation from donors and responsible private sector investors. This financial participation will help develop a shared agenda, enabling all groups to focus their inputs on the same basic set of objectives in the sector within the framework of enhanced NFPs. The disbursement of these funds will be handled in parallel—no centralized fund is proposed—using coordinated programs to support countries in realizing their NFPs. Financing partners already have indicated their willingness to work together both analytically and financially to realize these objectives.

The advantage of this arrangement is that participating donors would be relying on a single strategic approach: one that would emulate a grant-and-loan-based instrument to client countries. This approach should lead to greater focus and cohesiveness in investment strategy. All stakeholders will benefit from this approach, because it is based on developing assistance programs that focus on priority improvements, with a strong emphasis on necessary extensive sector analysis and stakeholder consultation. It will enable the application of the principles of selectivity and alignment across the activities of a group of participating financiers, rather than those of the Bank alone. It will improve the focus of interventions and lower reputational and institutional risk in the sector for all involved by ensuring more participation by stakeholders early in the commitment cycle. In addition:

- The Bank will gain leveraging impact for its financing, and other donors will gain the impact of coordinated policy reform and strong linkage of sector issues with broader economic frameworks.
- Opportunities will increase for private sector participation in responsible forest investments from the reduced risk expected to result from the improved policy and enabling environment and the more unified stakeholder approach to the sector.
- Clients will gain from efficiencies and streamlining in the sector programs they are planning and expected to implement.

Supporting and Strengthening the World Bank/WWF Alliance

The Bank has been supporting the World Bank/WWF Alliance through trust funds, the Development Grant Facility, and some Bank budget funds raised through the ESSD Network. It is necessary to ensure that the trust fund support continues: some commitments to the Alliance are time bound and will terminate at the end of FY02. There also needs to be adequate support in ESSD to manage this Alliance and in the Regions to continue support for operational staff to dedicate some of their time to act as Regional coordinators. Some prospects for expanding this initiative to focus more on the use of forest conservation and management for sustainable poverty reduction, to involve a broader group of NGO partners, and to involve the GEF more directly are under consideration by the Alliance Steering Committee.

Increasing Involvement with Private Sector Partners

For forest management to improve, responsible private investors who are willing to support SFM need to be brought into the sector and logging enterprises that participate in destructive and sometimes rogue and illegal forest operations need to be shut down.

If governments, the Bank, and its partners work together to develop a positive enabling environment for long-term and sustainable private sector investments in natural resources, responsible and environmentally conscious investors that are interested in supporting SFM and conservation can be brought into the sector. The potential to attract such investors is likely to increase as mechanisms are developed for investments in forest-based carbon through which afforestation, reforestation, and preserving standing natural forests can be made more financially attractive.

At present, the Bank's main interaction with the private sector on a multilateral basis is the CEOs Forum, in which private enterprises meet together with leading NGOs and representatives from other organizations. As in the case with the Africa Working Group of the CEOs Forum, collective action can bring about codes of conduct that better meet international standards of forest management. Working together through dialogue, third-party oversight, and encouragement of responsible institutional investments can change the landscape of private investment in production forests from nonsustainable practices to sustainable management and conservation. Indeed, a major priority of this revised Forest Strategy is to allow the Bank to play a significant role in helping what are all too rare examples of good forest management practice become private sector norms rather than private sector oddities. Abandoning the partnership with the private sector can lead only to the continuation of the current widespread patterns of economically inequitable, socially destructive, and environmentally irresponsible exploitation of the world's forests.

The IFC has many individual investments in various private sector companies. Drawing on these contacts will open another avenue by which the Bank Group could expand its involvement with the private sector. Maintaining the CEOs Forum, contacting potential new members, and expanding efforts through the Forum to generate specific field investments will require additional budgetary support for the Forum and increasingly shared responsibility.

SELECTIVITY

Given the limits on both Bank resources and capacity, it will be necessary to be selective as to where the efforts to increase Bank engagement in forests are to be applied and also to prioritize the specific forms of involvement to gain the most impact.

Alignment and Selecting Countries for Focus

The task of building engagement will consist of bringing selected candidates from among these potential investments through the CAS and PRSP processes, then into the project identification and concept development processes, and on into the Regional pipelines.

The essential items needed to realize a selected program of investments from this range of possibilities are the generation of Bank and borrower commitment to increased investments and partnerships with donors and others to create blended financing and attractive rates of return. If a decision is made to proceed with the strategy proposed in this document to rebuild Bank engagement in forests, the first step will be to select a group of client *focus* countries in which the Bank is (or will be) engaged in developing CASs, PRSPs, and/or significant natural resources or adjustment operations and in which forests are important in terms of both the environment and poverty reduction. Approximately 15 countries in this category will be identified, and initial evaluation of the prospects made. As indicated by the evaluation, six to eight countries will be identified to receive more intensive and direct support.

Developing the Focus on More Flexible Lending

One of the medium-term implementation objectives of the Forest Strategy is to move toward more flexible programmatic lending to assist borrowers to implement their forest programs with lower transaction costs. Under this strategy, the Bank will adopt a different approach depending on the readiness of countries to proceed with their enhanced NFPs (table 4.1). In some countries, the Bank will be able to participate only in country dialogue and to do initial sector work. At the other extreme, lending will be low cost with heavy reliance on government commitment to implement agreed forest programs. Facilitated by increased ESW, this approach is expected to lower lending costs and improve loan effectiveness in the forest sector.

Once the knowledge base has been built through sector work, project experience, and learning through monitoring, the Bank can proceed to more

Table 4.1 Appropriate Forest Investments

Category of client forest situation	Type of involvement	Cost/implementation implications
Category 1: Completed (or well advanced) enhanced National Forest Programs (ENFPs); demonstrated capacity to implement forest programs; and adequate policy and institutions, including fiduciary oversight and good governance.	Adopt longer-term, more flexible approach to support the sector. Continue adjustment lending and use guarantees where needed.	Bank: Lower budgetary costs; 0.75 to 1.0 percent of funds lent as preparation and supervision costs. Client: Lower risk status for private sector borrowing; higher leveraging of grant funds for protection and related purposes.
Category 2: Government and other major stakeholders committed to (or in early stages of involvement in) ENFPs or equivalent national sector strategy development; policy and institutional reforms and strengthening still needed.	Major participation in ENFP development. Follow-up investments for scaling up piloted approaches to national scale via project investment lending and adaptable program loans. Intensification of donor technical assistance in field operations and monitoring and evaluation. Possible use of sector adjustment instruments and guarantee mechanisms to stimulate scaling up of appropriate investments in the sector.	Bank: Medium budgetary cost: 1 to 2 percent of funds lent (depending on proportion of sector lending adjustment present in portfolio). Client: Reduced risk perceived by potential private sector lenders; higher prospects of Bank and other institutional loans being successful.
Category 3: Significant interest among stakeholders and reformist elements in government in forest sector management and development; major policy and institutional issues unresolved.	ESW and other investment in dialogue and sector analysis, investment in institution building and piloting feasible approaches to management, protection and improved participatory approaches (via learning and innovation loans). Possibly some incorporation of basic policy reforms in planned structural adjustment operations.	Bank: Higher cost, 2 percent-plus on lending amount due to continued up-front dialogue/ESW costs. Client: Limited exposure to lending in relatively poor implementation environment; opportunity to build a strategic approach and a strong constituency for reform.
Category 4: Government commitment and awareness on major sector issues weak or not present; other stakeholders poorly organized or in major conflict; major economic and social difficulties present.	ESW and other inputs to support initiation of dialogue and basic sector analysis; outreach and consultation to build sector constituency groups in-country. Strategic support through IFC and the World Bank/WWF Forest Alliance for NGOs and catalytic private or community operators prepared to meet international norms of good forest management practice.	Bank: Limited lending to governments at this stage; limited inputs into basic sector analysis, dialogue/watching brief activities. Client: Basic awareness building, initial stakeholder contacts.

ESW, economic and sector work; IFC, International Finance Corporation; WWF, World Wide Fund for Nature; NGO, nongovernmental organization.

flexible forms of lending at lower transaction costs. Table 4.1 indicates how this progression will take place and how the cost of lending will change with each step of increase in country ownership and commitment.

EXPECTATIONS AND REALITIES: RISKS AND MONITORING PROGRESS

A primary message of the consultations done for this strategy is that, to make a difference in the forest sector, the Bank must become more engaged.

The ultimate beneficiary of the Bank's renewed engagement in the forest sector through this strategy will be the poor and the global environmental commons. In the earlier of the initial five years of the strategy's implementation, the focus will be primarily analytical work and implementing the financial partnerships with donors, NGOs, and the private sector in focus countries willing to progress on forest issues.

Institutional Realities

It is acknowledged here that the tasks involved in the three pillars of strategy discussed in this paper are very large, that the potential outcomes for forests and forest-dependent people and countries with large forest resources are even larger (see the Broad Outcomes section below), and that these large expectations could cause some reputational risk for the Bank (this specific issue of risk is discussed under the Risks section below). The reality is that any feasible program for the Bank to reengage in the forest sector must begin modestly, recognizing that time and effort will be needed to build both the institutional capacity and the expanded demand from borrowers for Bank involvement. In its early stages, reengagement in the forest sector must be based on analysis and consultative work, linked to Bank and NFP priorities, and supported by partnership arrangements. Later developments will depend on outcomes of these activities and will proceed in the directions indicated by country commitments.

This caution is due largely to the reality that a reengagement in this sector will take time to build from the present low human and technical resource base available in the Bank. At the time of writing, the Bank had 25 operational staff who could be classified as being engaged in forest work for more than 50 percent of their time. Of these, approximately one-third are located in the field. A further five staff members work in the ESSD Forests Team, and the equivalent of three more are in the Development Economics Unit and WBI. More than half of these people are regular staff; the remainder are appointed on various consultancy contracts, and some are in the process of conversion to regular status. A further 12 staff work more specifically on biodiversity and conservation issues, which overlap forest areas to a significant extent.

Eventually, an increase in resources focused on forests will be essential. At the time of writing, the Environmental and Rural Sector Boards in the Bank had initiated a program to hire more natural resource specialists in the Bank, and it is significant that the Regions have requested that an emphasis be given to hiring staff qualified in forest areas. The time frame over which increased engagement in the sector occurs, and the nature of that engagement, will depend on specific Regional and Country circumstances. The Bank's commitment to reengage in the sector will itself raise the confidence of other donors, NGOs, and private sector investors that the Bank is willing to implement its Forest Strategy. Bank engagement will follow its comparative advantage in policy and institutional strengthening and scaling up successful programs of forest management and conservation.

Some Potential Broad Outcomes, in Perspective

It is useful to consider the magnitude of impacts that could result from an effective international program to address the Rio Forest principles, and subsequent intergovernmental agreements on what is needed in forests, to gain some appreciation of the importance of the task. On the basis of some previous commitments, and figures on the scale and magnitude of forests' potential contribution to development and environmental protection discussed earlier in this paper, it is reasonable to suggest that a significant and well-directed effort among donors, NGOs, the private sector, borrower country governments, and stakeholders could aim for the following large goals over the next 5 to 10 years:

Poverty. Improve the livelihoods of 500 million people, most of whom are poor and dependent on forest and tree resources, primarily through community forest management and development of agroforestry.

This number, while large, is less than half the number of poor people estimated to be dependent to some significant extent on forests for their livelihood.

Governance. Strengthen the institutional capacity to reduce the losses from illegal logging by US$5 billion per year and improve the management of forest concessions to increase government revenues by US$2.5 billion per year.

- As noted in chapter 1 of this paper, the estimated losses from failure to collect appropriate royalties and taxes from legal forest operations is costing governments about US$5 billion annually. Illegal operations probably cost them a further

US$10 billion in lost revenues. Recovery of half the amounts currently lost through improvements in the capacity of governments and other stakeholders to collect revenues, raise more reliable sources of financing for forest operations, and control illegal operations would represent a significant achievement, if not eliminating the problem entirely.

- An example from recent Bank experience cited in the current *World Development Report 2000/2001: Attacking Poverty* relates to significant reforms in governance and revenue collection that have occurred in Cameroon, through a cooperative effort over a number of years between the government and the Bank, in the context of structural adjustment operations. While the process is not yet complete, it has led to significant gains in transparency, revenue yield, participation by communities, and compliance with proper forest management procedures.

Protection and conservation. Bring 50 million ha of forests into new Protected Areas and improve the management of 50 million ha of currently Protected Areas.

These protection outcomes derive principally from commitments governments themselves have made in recent years, finalized into "stretch" targets as outlined by President Wolfensohn at the UNGASS meeting in 1997. There has in fact been significant progress with these targets, and it is likely they will actually be exceeded within the original 2005 time frame set by the president.

Sustainable forest management. Bring 200 million ha of global forests under SFM that is independently verified and certified. This "stretch" target was also one of the goals enunciated by the Bank president at the 1997 UNGASS meeting.

- Progress with this objective is likely to be slower than anticipated in 1997. However, the area of forest under certification has expanded exponentially since then, from virtually nil to about 27 million ha worldwide. About 9 million ha of this total is in Bank borrower countries, of which some 3 million is in tropical forest. Given that there are on the order of 600 million to 800 million ha of natural forest that already are, or will soon be, under some form of contractual or informal intent for production activities in the Bank's borrower countries, it would seem reasonable to aim at bringing 10 to 15 percent of this under improved standards of forest management within a 10-year period. Fulfilling this aim, along with continued rapid progress in developed country certification, would achieve this certification target.

The reduction of poverty through mechanisms that simultaneously improve the global environmental commons benefits all stakeholders. However, it is reiterated here that to achieve these goals will require the *concerted and coordinated efforts of all stakeholders.* This proposed strategy indicates a possible path for the Bank to play a significant part in that (albeit with limited resources) through a catalytic approach: building partnerships and knowledge and increasing the blended financial flows to developing countries needed to realize the potential contributions of forests to sustainable conservation, development, and poverty reduction.

Risks

The most immediately obvious risk for the Bank in adopting the new strategy proposed in this paper was the necessity to modify the 1993 Forest Policy, OP 4.36, a move that was deemed necessary to achieve the aims of this strategy. This risk was especially high if the replacement of the present broad-based ban on Bank financing of logging activities in all of the primary tropical moist rain forest with more closely targeted provisions were to be seen as simply "opening the floodgates" to Bank support for destructive and unsustainable logging operations in these forests.

- This issue has been extensively discussed in chapter 2 of this paper. It is made clear in that discussion that the intention of modifying the policy in this area was to ensure that the Bank becomes an effective player in the management of forests in an appropriate manner and utilizes various new approaches that will allow this to happen. There has been no financing of unsustainable logging by the Bank in Regions where the ban in the 1993 OP does not apply (since it applies only in primary tropical moist forests). There is in fact no indication of borrower demand for Bank financing of such activities.
- Bank support for commercial harvesting activities is, in any event, restricted in the new policy to situations where there will be independent validation of agreed standards in forest operations.

The new OP utilizes new developments in independent assessment and certification of forest operations to ensure that any operations supported in any way by the Bank will be significant improvements over the norm, in any given situation. The approach will also provide an ongoing mechanism by which performance in the objective of improving the social, environmental, and sustainable development objectives of forest operations are in fact being met. This will have a strong collateral effect on the compliance with *other* safeguard policies, since in the area of natural habitats, indigenous people, and cultural sites, the requirements of the Forest Policy specify in a forest context the requirements of these other safeguards, and therefore would incorporate these in criteria and indicators for forest operations under some form of independent assessment or certification.

- It is made clear in this strategy that the large majority of stakeholders who were consulted in all regions of the world on the issue of the Bank's policy and its approach were strongly in favor of the Bank, adopting the policy measures now in the new OP (included on the CD), as an essential part of its reengagement in the sector, which many saw as long overdue.

A major risk for the Bank in the approach outlined in this strategy paper is in the large gap that many will perceive between the tasks needed to achieve significant progress with the three pillars outlined in chapter 1 as well as with the international community's Forest Policy objectives, on the one hand, and the fairly cautious and modest measures proposed for the Bank to initiate its reengagement in forests, on the other hand.

- It is essential that the Bank should acknowledge not only the size and urgency of the task in forests, *but also* the limitations that apply to its own capacity to deal with these, at least in the immediate and near-term future. Even when combined with the most optimistic projections of what other multilateral banks and donor agencies might bring to this sector, the total flow of resources will continue to be dwarfed by the flow of private sector funds into the sector (much of which is presently focused on inappropriate and unsustainable use of forests), and also the flow of all funds into other parts of the economies of important forest countries—which has potentially large impacts on forests. This is why the approach outlined in this paper emphasizes the *catalytic* role of the Bank and other donor partners in the forest sector: the objectives must be to promote an enabling environment that will encourage the right kind of investment to ensure that the three pillars are addressed effectively; to utilize the Bank's influence to draw the right kind of private sector financing into the sector; and to increase its own abilities to incorporate forest and forest peoples issues effectively into its own broader-based investments with its borrower countries.

A related risk is the possibility that insufficient interest and commitment will arise in borrower countries, as a result of the inputs proposed here, to generate the major changes and reforms necessary and the demand for investments needed to achieve them. Two basic approaches to mitigation of this risk are embedded in the overall approach proposed:

- The nature of the ESW proposed is directly focused on this issue: the Bank and other partners will work directly with borrower governments and other local stakeholders to identify the real economic and environmental opportunities that forests offer, the costs of developing these opportunities, *and* the costs of ignoring them. This will help ensure that the real situation is clearly understood by all major stakeholders and allow all to focus on the priorities and realities involved in managing and protecting the forests and the interest of forest people.
- The catalytic nature of the involvement outlined above, as it applies to the private sector, will have particular application to mitigating this risk. The Bank will utilize its convening power to bring new financiers into the sector in borrower countries, including external and large-scale "green" investors, and local investors, who to date have been reluctant to enter this sector given its perceived problems of governance and nontransparency in many cases.

A parallel risk to that of commitment by borrowers and other stakeholders exists internally in the Bank among managers and staff within the Bank, as a result of the "chilling effect" on direct involvement in forests resulting from the 1993 Forest Policy that OED has identified, and the perceived high transaction costs of dealing with this sector, which is connected to the need to comply with other safeguard policies. The risk is that the reluctance to deal with the more difficult aspects of the sector created by these constraints may be too strong and entrenched to be overcome by the measures proposed in this strategy. First, it needs to be acknowledged here that the overall package of safeguard measures the Bank applies to investments and other programs that involve natural forests are likely to lead to more effective and sustainable projects. Equally, however, there is little doubt that, under present conditions, meeting the demands of these safeguards in project design and implementation can be onerous and expensive. This is due in large part to the nature of forests, which are a complex of competing goods and services, often further complicated by unresolved and frequently contentious issues of access, ownership, and rights among various groupings in society. The approach outlined in this paper will provide some effective mitigation of the problem of continued risk aversion to direct involvement in forests within the Bank in three ways:

- The new OP underpins a new, proactive attitude to engagement in natural forests by the Bank, and, if adopted, will send a clear signal to both Bank staff and borrowers that this is the case. This will directly address the "chilling effect."
- As noted under the discussion of the risks associated with adoption of this new policy, an important collateral effect of the independent assessment and certification measures proposed as part of the new OP will also assist in identifying compliance issues in the other important safeguards that apply to forest protection and use. This will provide some assurance to staff and managers that a competent and continuous process of assessment of compliance with important safeguards is built into the operation of a project and will provide timely warning of possible problems, so that these can be dealt with effectively.
- While no specific lending targets are proposed in this strategy, it is recommended (in the Business Strategy attachment) that when interest, commitment, and demand for further engagement in the Bank are built through the enhanced ESW program proposed, the Bank will need to consider the use of corporate funds, if necessary, to reduce the incremental cost burden that preparation and supervision of projects based on forests incur for Country Departments. While it will remain a decision for Bank management at the time whether and how this assistance will be provided, it has some precedent in the Bank, in the form of central funding of global public goods initiatives.

There is, as always, a significant risk for the Bank inherent in entering into new partnerships. The essential risks are (a) that accountabilities and roles for all parties to the arrangement will not be clearly enunciated and negotiated at the outset, leaving the likelihood that the activities and outcomes may not be agreed upon later, or that the Bank, as host organization, will need to absorb more responsibility than it can manage; (b) that parties' priorities and capacities may change during the course of the partnership program, leading to the possibility that they will not continue to meet their agreed obligations and responsibilities; and (c) that the parties to the partnership may have widely differing interpretations of its role and function.

Two issues remain to be addressed in this discussion of risk:

First, an effective communication strategy will be essential, to ensure that the development and implementation of this strategy are clearly enunciated and understood, and the constraints and realities are recognized from the outset. As a first measure, the posting of this strategy and the draft revision of the OP for public review will be accompanied by a comprehensive questions-and-answers paper, which will explain in depth the reasons for adoption of certain measures and decisions that may prove controversial, and which will also explain the reasons for adoption of a gradual scaling up of reengagement and a catalytic approach, as discussed above. A second important element in the communication strategy will be the proposed EAG for implementation of this strategy: in addition to providing independent advice on implementation issues, this body will serve as a conduit to interest groups for information on the status and effectiveness of the program, and in this way will serve a valuable transparency role.

Second, when evaluating the risks and mitigating measures as discussed above, it must be borne in mind that there would also be major risks for the Bank in maintaining the status quo in its previous Forest Strategy and Policy:

- As noted above, certain stakeholders will regard any variation of the broad-based ban on support for logging activities in the tropical moist forest as an "opening of the floodgates" (and in fact this group will press for an extension of this ban to cover all old growth forest in nontropical areas, as a minimum condition). However, the 1993 Policy, even if extended in this particular way, was criticized by the majority of stakeholders and interested parties outside the Bank as being a significant factor in the Bank's perceived withdrawal in recent years from major areas of the sector. In these circumstances, the best approach to managing this risk is not to adhere to the status quo but to develop a proactive policy that engages stakeholders with the Bank in defining acceptable forest outcomes and then monitoring activities to ensure that they are achieved. It is certain that a much larger group of stakeholders, including some important international environmental NGOs, agree with this approach and will regard any move by the Bank to circumscribe or minimize its proactive reengagement in this sector as a major failure of will.

- Certainly, this would prejudice the possibilities of a growth in borrower demand for Bank involvement in forests, since the interpretation would be that the Bank continues to have limited interest in this sector.
- The strategy calls for an enhanced commitment to ESW and proposes a major partnership to assist in implementing that. Among other things, this will allow forest issues and impacts that might be of relevance to larger Bank programs being considered to be considered in a timely manner. Without this, there are possibilities that Bank nonforest sector activities might inadvertently cause deleterious impacts on forests and forest peoples that could have been avoided without prejudicing the basic aims of the investment, had the right offsetting measures been designed into the process.
- The partnerships proposed for inception or expansion in this strategy will build linkages among the Bank and the private sector, the donor community, and major NGOs. In each case, this will lead directly to benefits from coordination and shared knowledge that are not present (or are present to a more limited extent) in present arrangements. Given the high levels of activity from all quarters in the forest sectors of Bank borrower countries, the partnerships proposed will certainly lower the risk of overlapping initiatives; poor sequencing of respective donor, NGO, and private sector initiatives; and discontinuities in the assistance provided that has characterized this sector in recent years.

None of these approaches will be easy to implement, and all will require an ongoing commitment by the management and the Board of the Bank. However, there is little alternative to this: so far as the very visible challenges now presented by forests and forest-dependent people to the Bank are concerned, attempts to preserve the status quo will provide no refuge from political controversy and criticism.

Monitoring the Bank's Role

As will be apparent from the above discussion, the Bank can expect to encounter major difficulties inherent in attempting to contribute effectively to large tasks and objectives with limited resources by relying on catalytic activities and strong partnerships—but it has little alternative but to embark on this approach. For a decentralized institution such as the Bank, joint engagement is even more difficult when global priorities are involved. However, joint engagement is essential to success in the forest sector and other sectors where effective engagement of the Bank and other partners is indispensable to achieving the MDGs.

Monitoring and reviewing progress will therefore be an important activity for this strategy, since it will make it possible to assess whether measures and activities are correctly focused and adequate. Monitoring implementation of the strategies for the increased engagement proposed for the Bank will be based on the following performance indicators:

Developing demand. The Bank will monitor the effectiveness of incremental analytical and consultative activities in the focus countries. Criteria will comprise the alignment of these activities with the major pillars of engagement put forward in this strategy and with NFP activities and objectives; the establishment of working partnerships at the country level with other donors and major local stakeholders; and the degree of ownership and consensus on major elements of reform in the sector, including the development of interest and participation by major agencies of government in the agenda.

Building engagement. The test of this component will come later in the program in the form of initiating the preparation of investments that have resulted from the ESW activities. In addition to normal Bank quality-at-entry and performance indicators, an important additional criterion will be the assessment of the degree of parallel financing or cofinancing by donor partners and the private sector, which emerges as a result of collaborative activities. Base lining at the beginning of the program will be an important measure to allow assessment of this factor.

Impact on forest outcomes of broader involvement. It also will be necessary to evaluate the impact of the knowledge and analysis generated through this strategy on Bank activities originating outside the forest sector: the identification of critical forest issues; their incorporation in cross-sectoral, adjustment, and broad programmatic lending; and the inclusion of forest topics in broader ESW and CAS development work programs.

Selectivity and sequencing. Effective disbursement and implementation of the incremental activities proposed, in the time outlined, will be a major factor. The balance of activity types and progress toward lowering overall costs of doing business in the sector through a movement of country situations being addressed up through the hierarchy of engagement, as illustrated in table 4.1, will be a major factor for evaluation.

Mid-term evaluation. Provision for a specific mid-term review is included in the Business Plan for this strategy. The objectives and terms of reference for this will need to be carefully designed to concentrate evaluation on ESW as well as whatever investment implementation is in progress. The mid-term review should be carried out by OED, and the proposed EAG that is intended to support implementation of the new strategy would be intensively engaged in this exercise as well.

External Advisory Group

External assessment will be built into the evaluation process, continuing the approach used during the preparation of the FSSP. Stakeholders will be consulted at every stage. Specifically, an ad hoc EAG will be formed, utilizing the former TAG formed to assist the Bank with development of the new strategy and policy as a roster for selection on as-needs basis. Small teams (8 to 10 members) would be formed for specific tasks, comprising individuals from major stakeholder groups (client governments, civil society groups, academic and private sector interests, and major donor partners) represented on the former TAG. The EAG process would be formulated in FY03, in time to allow selected group(s) to assess design, relevance, and progress of the activities for implementation outlined below, with a specific objective of advising on implementation matters related to the Bank's OP and the overall aims of the strategy as set forward in the FSPP. In addition, the EAG will regularly advise the president of the World Bank regarding its view of the status of the forest program.

NOTES

1. The Millennium Development Goals affirm the international community's commitment to free individuals from the dehumanizing conditions of abject poverty. The United Nations (UN) announced the goals in September 2001 following extensive consultations among the UN Secretariat, International Monetary Fund (IMF), Organisation for Economic Co-operation and Development, and the World Bank. These targets and indicators represent the harmonization of the International Development Goals agreed at various global conferences and summits of the preceding decade with the UN Millennium Declaration adopted in September 2001 by 147 heads of state. www.imf.org/external/np/sec/nb/2001/nb0190.htm
2. An example is the recent study by Schneider and others (2002 [English], 2000 [Portuguese]), "Sustainable Amazon: Limitations and Opportunities for Rural Development." This study shows that 83 percent of the area of the Amazon is unsuitable for agriculture and ranching and that continuation of these activities in forests will result in extremely low returns from this type of land use, as well as permanent loss of the forest areas.
3. See World Bank/WWF Alliance for Forest Conservation and Sustainable Use 1998. The Alliance believes that a common set of principles should underscore any standard for improving the management of both natural and planted forests. These principles include compliance with all relevant laws, documented tenure and use rights, respect for indigenous peoples' rights, respect for community relations and workers' rights, encouragement of multiple benefits from the forest, containment of the environmental impacts of forest use, rigorous forest management planning, active monitoring and assessment, and the maintenance of critical natural forest areas. The Alliance believes that credible certification systems must be consistent with the following criteria:

   - Institutionally and politically adapted to local conditions
   - Goal oriented and effective in reaching objectives
   - Acceptable to all involved parties
   - Based on performance standards defined at the national level that are compatible with generally accepted principles of sustainable forest management
   - Based on objective and measurable criteria
   - Based on reliable and independent assessment
   - Credible to major stakeholder groups (including consumers, producers, and conservation NGOs)
   - Certification decisions free of conflicts of interest from parties with vested interests
   - Cost-effective
   - Transparent
   - Equitable access to all countries

4. See discussion of the conditions for World Bank involvement regarding the preservation of intact forest areas in "The Forest Sector." (World Bank 1991e, pp. 19–20, 65–66.)
5. A 1989 ITTO study found that less than 1 percent of commercial operations in tropical forests globally met reasonable operational standards for sustainable forest management. See Poore and others 1989.
6. To illustrate: In 1998 the government of Brazil announced its intention to develop a mosaic of conservation areas in the Amazon within which some 50 million ha of national forests under sustainable harvesting and management operations will form a buffer around a network of national parks and

indigenous reserves covering approximately two-thirds of the Amazon forest region.

7. Mitigation is defined here as human intervention to reduce the sources of greenhouse gases or enhance their sinks.
8. Adaptation is defined as the ability to adjust to climate change (including climate variability and extremes), to moderate potential damages, to take advantage of opportunities, or to cope with the consequences.
9. The two major flexible mechanisms under the Kyoto Protocol are Joint Implementation (JI), which deals with carbon emission trade in the "Appendix 1" countries, and the CDM, which deals with carbon emission trade between Appendix 1 countries and developing countries. The role of forests and forestry still needs to be specified in the CDM framework.
10. The CPF is composed of all former members of the Interagency Task Force on Forests (ITFF), along with some other organizations. Current members include the United Nations Framework Convention on Biological Diversity (CBD), CIFOR, FAO, ICRAF, ITTO, IUCN, United Nations Convention to Combat Desertification, United Nations Department of Economic and Social Affairs, UNDP, United Nations Environment Programme (UNEP), UNFCCC, and the World Bank.
11. PROFOR was established in 1997 in response to the proposals for action of the IPF to promote SFM and related public and private sector partnerships and thus to enhance forests' contribution to sustainable livelihoods. At the national level, PROFOR's collaborative work supports countries in developing and implementing Forest Policy planning and implementation processes that more effectively address local needs and national priorities and reflect the internationally agreed principles for NFPs. It also promotes the use of NFPs as a basis to enhance cooperation in the forest sector. At the international level, PROFOR generates and promotes knowledge of NFP processes and critical thematic issues related to NFPs—notably, livelihoods, governance, and financing strategies—among client countries and international organizations.
12. Even in the absence of this agreement, carbon trading is taking place using developing country forests as sinks. This activity has become policy for some corporations ahead of any formal international agreement on mechanisms.
13. It is axiomatic that, to succeed, a strategy for building the Bank's engagement and effectiveness in forest outcomes will need to be:

    - Implemented through country teams and programs, beginning with the clear reflection of intentions and priorities in CASs, PRSPs, and business plans.
    - Activated under the management of Regional Vice Presidents, who will hold the responsibility to link outcomes under the newly adopted Bank global issues with country programs.
    - Demand driven with sufficient attention and resources devoted to dialogue and incentives to attract the interest of borrower governments, other major stakeholders, and Bank Country Departments to invest in the forest sector.

BIBLIOGRAPHY

The word *processed* describes informally reproduced works that may not be commonly available through libraries.

African Regional Workshop. 1998. "Report on the African Regional Workshop on the Underlying Causes of Deforestation and Forest Degradation." Accra, Ghana.

Agarwal, Bina. 1997. "Gender, Environment, and Poverty Interlinks: Regional Variations and Temporal Shifts in Rural India, 1971–91." *World Development* 25(1):23–52.

Anderson, Dennis. 1987. "The Economics of Afforestation." World Bank Occasional Paper 1. Washington, D.C.

Bass Stephen, and M. Simula. 1999. "Independent Certification/Verification of Forest Management." Background paper prepared for the World Bank/WWF Alliance Workshop, Washington, D.C., November 9–10. International Institute for Environment and Development. Unpublished.

Binkley, C. S. 1999. "Forest in the Next Millennium: Challenges and Opportunities for the USDA Forest Service." Discussion Paper 99-15. Resources for the Future, Washington, D.C.

Bird, Pippa. 2000. "A Livelihoods Approach: Why Bother?" United Nations Development Programme, Programme on Forests. New York. Processed.

Blaser, Juergen, and James J. Douglas. 2000. "Issues and Implications for the Emerging Forest Policy and Strategy of the World Bank." Tropical Forest Update 10. International Tropical Timber Organization, Yokohama.

Bojö, Jan, and Stefano P. Pagiola. 2000. "Natural Resources Management." World Bank, Environment Department, Washington, D.C.

Boscolo, Marco, and Jeffrey R. Vincent. 1998. "Promoting Better Logging Practices in Tropical Forests: A Simulation Analysis of Alternative Regulations." World Bank, Development Research Group, Washington, D.C.

Bowles, Ian, Richard Rice, Russell Mittermeier, and Gustavo A. da Fonseca. 1998. "Logging and Tropical Forest Conservation." *Science* 280:1899–1900.

Bruner, A. G., R. E. Rice, and G. A. B. da Fonseca. 2000. "Effectiveness of Parks in Protecting Tropical Biodiversity." *Science* 291:125–28.

Campbell, J. Gabriel, and Alejandra Martin. 2000. "Financing the Global Benefits of Forests: The Bank's GEF Portfolio and the 1991 Forest Strategy." Report Number 20627. World Bank, Washington, D.C.

Carter, Jane. 1999. "Recent Experience in Collaborative Forest Management Approaches." World Bank, Washington, D.C.

Cernea, Michael M. 1989. "User Groups as Producers in Participatory Afforestation Strategies." World Bank Discussion Paper 70. Washington, D.C.

Chomitz, Kenneth M. 2000. "Evaluating Carbon Offsets from Forestry and Energy Projects." Policy Research Working Paper 2357. World Bank, Washington, D.C.

Chomitz, Kenneth M., and Kanta Kumari. 1998. "The Domestic Benefits of Tropical Forests: A Critical Review." *The World Bank Research Observer* 13(1):13–35.

Contreras-Hermosilla, Arnoldo. 1997. "The Cut and Run Course of Corruption in the Forestry Sector." *Journal of Forestry* 95(12):33–36.

Contreras-Hermosilla, Arnoldo, Aaron Zazueta, Syed Arif Husain, Kavita Gandhi, Alejandra Martin, and Uma J. Lele. 2000. "Preliminary Review of the Implementation of the Bank's 1991 Forest Strategy in the Latin America and Caribbean Region." World

Bank, Operations Evaluation Department, Washington, D.C.

Costanza, Robert, Ralph d'Arget, Rudolf de Groot, Stephen Farber, Monica Grasso, Bruce Hannon, Karin Limburg, Shahid Naeem, Robert V. O'Neill, Jose Paruelo, Robert G. Raskin, Paul Sutton, and Marjan van den Belt. 1997. "The Value of the World's Ecosystem Services and Natural Capital." *Nature* 15 (May).

Crooks, Robert, William B. Magrath, Glenn Morgan, and Susan Shen. 1999. "Natural Resource Management: A Strategic Framework for East Asia and the Pacific." EAP Working Paper. World Bank, East Asia and Pacific Department. Washington, D.C.

Department of the Environment. 1991. *Interpretation of Major Accidents to the Environment for the Purposes of the CIMAH Regulations.* London.

DFID (Department for International Development). 2000. "Achieving Sustainability: Poverty Elimination and the Environment." Target Strategy Paper. London.

Doolette, J. B., and William B. Magrath. 1990. "Watershed Development in Asia: Strategies and Technologies." World Bank Technical Paper 127. Washington, D.C.

Essama Nssah, Boniface, and James Gockowski. 1999. "Forest Sector Development in a Difficult Political Economy: An Evaluation of Cameroon's Forest Development and World Bank Assistance." World Bank, Operations Evaluation Department, Washington, D.C.

FAO (United Nations Food and Agriculture Organization). 1999. "Sustainable Forest Management Issue Paper." Rome, Italy.

———. 2000. *Global Forest Resource Assessment.* Rome, Italy.

FAO and World Bank. 1999. "Technical Consultation on Management of the Forest Estate: Issues and Opportunities for International Action by the World Bank and FAO." Proceedings, April 28–29, 1999. Rome, Italy.

Gilmour, D. A., and R. J. Fisher. 1991. *Villagers, Forests and Foresters: The Philosophy, Process and Practice of Community Forestry in Nepal.* Katmandu, Nepal: Sahayogi Press.

Goodland, Robert, and Jean R. Mercier. 1999. "The Evolution of Environmental Assessment in the World Bank: From Approval to Results." Environment Department Paper 67. World Bank, Washington, D.C.

Gregersen, Hans. 1998. "Evolving Forest Paradigms: Implications for World Bank Policy and Action." College of Natural Resources, University of Minnesota.

Guerguieva, Anna, and Kirk E. Hamilton. 2000. *A Brief Retrospective on Adjustment Lending and the Environment.* Washington, D.C.: World Bank, Environment Department.

Hamilton, Kirk E. 2000. "Mainstreaming Environment in Country Assistance Strategies." World Bank, Environment Department, Washington, D.C.

Hassan, Hassan, and H. E. Dregne. 1997. "Natural Habitats and Ecosystems Management in Drylands: An Overview." Environment Department Paper 51. World Bank, Washington, D.C.

Hobley, M., J. Y. Campbell, and A. Bhatia. 1996. "Community Forestry in Nepal: Learning from Each Other." MNR96/3. Katmandu, Nepal.

Hyde, William F., and Roger Sedjo. 1992. "Managing Tropical Forests: Reflections on the Rent Distribution Discussion." *Land Economics* 68(3):343–50.

ICRAF (World Agroforestry Center). 1998. "More People; More Trees." Nairobi, Kenya.

Imazon. 1999. "Hitting the Target: Timber Consumption in the Brazilian Domestic Market and Promotion of Forest Certification." FOE Programa Amazonia, São Paulo; Belem, PA, Imazon; Piracicaba, SP, Imaflora.

IPCC (Intergovernmental Panel on Climate Change). 2000. *Cross Cutting Issues Guidance Papers.* Tokyo, Japan: Global Industrial and Social Progress Research Institute.

———. 2000. *Land Use, Land-Use Change, and Forestry.* A Special Report of the IPCC. Cambridge: Cambridge University Press.

IUCN and the World Conservation Monitoring Centre. 1994. "Guidelines for Protected Area Management Categories." IUCN (The World Conservation Union), Gland, Switzerland.

Jodha, N. S. 1992. "Common Property Resources: A Missing Dimension of Development Strategies." World Bank Discussion Paper 169. Washington, D.C.

Kaimowitz, David. Undated. "Protected Areas and Tropical Logging from a Political Economy Perspective." Center for International Forestry Research, Bogor, Indonesia. Processed.

Kaimowitz, David, and Arild Angelsen. 1998. "Economic Models of Tropical Deforestation, a Review." Center for International Forestry Research, Bogor, Indonesia.

———. 1999. *The World Bank and Non-Sector Policies That Affect Forests.* Jakarta, Indonesia: Center for International Forestry Research.

Kishor, Nalin M., and Luis F. Constantino. 1993. "Forest Management and Competing Land Uses: An Economic Analysis for Costa Rica." LATEN Dissemination Note 7. World Bank, Latin America and Caribbean Region, Washington, D.C.

Kumar, Nalini, and Uma J. Lele. 2000a. "Preliminary Review of the Implementation of the Bank's 1991 Forest Strategy and Its Implementation in the South Asia Region." World Bank, Operations Evaluation Department, Washington, D.C.

Kumar, Nalini, Ridley Nelson, Uma J. Lele, Boniface Essama Nssah, Kavita Gandhi, and Alejandra Martin. 2000b. "Preliminary Review of the Implementation of the Bank's 1991 Forest Strategy in the Africa Region." World Bank, Operations Evaluation Department, Washington, D.C.

Kumar, Nalini, Naresh Saxena, Yoginder Alagh, and Kinsuk Mitra. 2000c. *India: Alleviating Poverty through Forest Development.* Evaluation Country Case Studies.

Washington, D.C.: World Bank, Operations Evaluation Department.

Kumari, Kanta. 1994. "Sustainable Forest Management in Peninsular Malaysia: Towards a Total Economic Valuation Approach." PhD thesis. University of East Anglia, United Kingdom.

Lele, Uma J., Kinsuk Mitra, and O. N. Kaul. 1994. "Environment, Development and Poverty: A Report of the International Workshop on India's Forest Management and Ecological Revival." Center for International Forestry Research, Jakarta, Indonesia.

Lele, Uma J., Arnoldo Contreras-Hermosilla, and Kavita Gandhi. 2000a. "Preliminary Review of the Implementation of the Bank's 1991 Forest Strategy in the Europe and Central Asia Region." World Bank, Operations Evaluation Department, Washington, D.C.

Lele, Uma J., Saeed Rana, and Kavita Gandhi. 2000b. "Preliminary Review of the Implementation of the Bank's 1991 Forest Strategy in the East Asia and Pacific Region." World Bank, Operations Evaluation Department, Washington, D.C.

Lele, Uma J., Virgilio M. Viana, Adalberto Veríssimo, Stephen Vosti, Karin Perkins, and Syed Arif Husain. 2000c. "Forests in the Balance: Challenges of Conservation with Development." World Bank, Operations Evaluation Department, Washington, D.C.

Lele, Uma J., Boniface Essama Nssah, Madhur Gautam, and Saeed Rana. 2000d. "Limits of Environmental Adjustment for Long-Term Sustainable Growth." World Bank, Operations Evaluation Department, Washington, D.C.

Lele, Uma J., Nalini Kumar, and Kavita Gandhi. 2000e. "Forest-Dependent Poor, Institutional Change, and the World Bank's Role." World Bank, Operations Evaluation Department, Washington, D.C.

Lele, Uma J., Nalini Kumar, Syed Arif Husain, Aaron Zazueta, and Lauren Kelly. 2000f. *The World Bank Forest Strategy: Striking the Right Balance.* Washington, D.C.: World Bank.

Mahar, Dennis J., and Cecile E. H. Ducrot. 1998. "Land-Use Zoning on Tropical Frontiers: Emerging Lessons from Brazilian Amazon." Case Studies. World Bank Institute, Washington, D.C.

Malla, Sapana P. 1996. "Nepal Madhyasthata Samuha: Jalbire Women's Community Forestry Group." MNR 96/1. Katmandu, Nepal.

Margulis, Sergio, and Tonje Vetleseter. 1999. "Environmental Capacity Building: A Review of the World Bank's Portfolio." Environment Department Paper 68. World Bank, Washington, D.C.

Mayers, J., S. Bass, and D. J. Macqueen. 2002. "The Pyramid: A Diagnostic and Planning Tool for Good Forest Governance." Report prepared for World Bank/WWF Alliance. International Institute for Economic Development, London.

McCalla, A. F., and Wendy S. Ayres. 1997. *Rural Development: From Vision to Action.* Washington, D.C.: World Bank.

McCool, S. F., and D. N. Cole, eds. 1997. *Limits of Acceptable Change and Related Planning Processes: Progress and Future Directions.* U.S. Forest Service, Gen. Tech. Rep. INT-GTR-371, Ogden, Utah.

Palo, Matti, and Jussi Uusivuori, eds. 1999. *World Forests: Society and Environment.* Dordrecht/London/Boston: Kluwer Academic.

Pearce, David W. 1995a. *Blueprint for Capturing Global Environmental Values.* Blueprint 4. London: Earthscan Publications.

Pearce, David W. 1995b. "Global Environmental Value and the Tropical Forests: Demonstration and Capture." Prepared for Forestry and the Environment: Economic Perspectives II, Alberta, Canada, October 1994.

Pearce, David W., Francis E. Putz, and Jerome K. Vanclay. 1999. *A Sustainable Forest Future?* London: Natural Resources International and U.K. Department for International Development.

Picciotto, Robert. 1992. "Participatory Development: Myths and Dilemmas." Policy Research Working Papers WPS930. World Bank, Washington, D.C.

Poore, Duncan, Peter Burgess, John Palmer, Simon Rietbergen, and Timothy Synnott. 1989. *No Timber without Trees: Sustainability in the Tropical Forest.* London: Earthscan Publications.

Poore, Duncan, with Juergen Blaser, Eberhard F. Bruenig, Peter Burgess, Bruce Cabarle, David Cassells, James J. Douglas, Donald Gilmour, Patrick Hardcastle, Gary Hartshorn, David Kaimowitz, Nalin Kishor, Alf Leslie, John Palmer, Francis E. Putz, M. N. Salleh, Nigel Sizer, Timothy Synott, Frank Wadsworth, and Timothy Whitmore. 1998. "No Forest without Management: Sustaining Forest Ecosystems under Conditions of Uncertainty." Submitted to *Science.*

Prototype Carbon Fund. 2001. "Prototype Carbon Fund: A Public-Private Partnership to Mitigate Carbon Change. Annual Report 2001." Washington, D.C.

Putz, Francis E., Kent H. Redford, John G. Robinson, Robert Fimbel, and Geoffrey M. Blate. 2000. "Biodiversity Conservation in the Context of Tropical Forest Management 2000." Biodiversity Series. World Bank, Environment Department, Washington, D.C.

Rice, Richard, Cheri Sugal, and Ian Bowles. 1999. "Sustainable Forest Management: A Review of the Current Conventional Wisdom." Conservation International, Washington, D.C.

Rozelle, Scott, Jikun Huang, Syed Arif Husain, and Aaron Zazueta. 2000. *China: From Afforestation to Poverty Alleviation and Natural Forest Management.* Evaluation Country Case Study Series. Washington, D.C.: World Bank.

Schneider, Robert R. 1992. "Brazil: An Analysis of Environmental Problems in the Amazon." Vol. 1: Main Report. Report No. 9104-BR. World Bank, Washington, D.C.

Schneider, Robert R., Eugenio Arima, Adalberto Veríssimo, Paulo Barreto, and Carlos Souza Júnior. 2000. *Amazonia Sustentável: limitantes e oportunidades para o desenvolvimento rural.* Partnership Series 1. Brasilia: World Bank and Imazon.

Schneider, Robert R., Eugenio Arima, Adalberto Veríssimo, Paulo Barreto, and Carlos Souza, Jr. 2002. *Sustainable Amazon: Limitations and Opportunities for Rural Development.* Washington, D.C.: World Bank Technical Paper/Environment Series.

Seymour, Frances J., and Navroz K. Dubash, 2000. *The Right Conditions: The World Bank, Structural Adjustment, and Policy Reform.* Washington, D.C.: World Resources Institute.

Sharma, Narendra P., Simon Rietbergen, C. R. Heimo, and Jyoti Patel. 1994. "A Strategy for the Forest Sector in Sub-Saharan Africa." World Bank Technical Paper 251. Washington, D.C.

Shepherd, Gill, Michael Arnold, and Stephen Bass. 1999. "Forests and Sustainable Livelihoods." Draft. World Bank, Washington, D.C.

Shogren, Jason F., and John Tschirhart. 1999. "The Endangered Species Act at Twenty-Five." *Choices* (3d quarter).

Shrestha, K. B. 1996. "Nepal Madhyasthata Samuha Community Forestry in Nepal—An Overview of Conflicts." MNR 96/2. Katmandu, Nepal.

Shrestha, N. Kaji, Ghanendra Kafle, and Charla Britt. 1997. "Community Forest User Group Networking and Emergence of a Federation of Community Forestry Uses in Nepal." *Forests, Trees and People Newsletter* No. 32/33 February.

Simula, Markku. 1999. Independent Certification/ Verification of Forest Management." World Bank/ WWF Alliance. Washington, D.C.

Spears, John. 1998. "Sustainable Forest Management: An Evolving Goal." A position paper prepared for United Nations Development Programme's Global Programme, Washington, D.C.

Spilsbury, Michael J., and David Kaimowitz. Undated. "The Influence of Policy Research and Publications on Conventional Wisdom and Policies Affecting Forests." Center for International Forestry Research, Bogor, Indonesia.

Tomich, Thomas P., David E. Thomas, Y. Kusumanto, and Meine van Noordwijk. 1998. "Policy Research for Sustainable Upland Systems in Southeast Asia." International Centre for Research in Agroforestry, Nairobi, Kenya.

Tomich, Thomas P., Meine van Noordwijk, Stephen A. Vosti, and Julie Witcover. 1998. "Agricultural Development with Rainforest Conservation: Methods for Seeking Best-Bet Alternatives to Slash-and-Burn, with Applications to Brazil and Indonesia." *Agricultural Economics* 19:159–74.

Treeweek, Jo. 1999. *Ecological Impact Assessment.* London: Blackwell Science.

United Nations. 1997. "Report of the *Ad Hoc* Intergovernmental Panel on Forests on Its Fourth Session." Commission on Sustainable Development of the Economic and Social Council. E/CN.17/1997/12. New York.

———. 2000. "Economic and Social Council Resolution E/2000/L.32." Economic and Social Council, New York.

UNDP (United Nations Development Programme), UNEP (United Nations Environment Programme), and World Bank. 1994. *Global Environment Facility: Independent Evaluation of the Pilot Phase.* Washington, D.C.: World Bank.

Velozo, Ronnie de Camino, Olman Segura, Luis Guillermo Arias, and Isaac Perez. 2000. "Forest Strategy and the Evolution of Land Use: An Evaluation of Costa Rica's Forest Development and World Bank Assistance." World Bank, Operations Evaluation Department, Washington, D.C.

Vidaeus, Lars, and Gonzalo Castro. 2000. "The World Bank and the Global Environment: A Progress Report." World Bank, Washington, D.C.

Vosti, Stephen A., Julie Witcover, S. Oliveira, and M. D. Faminow. 1997. "Policy Issues in Agroforestry: Technology Adoption and Regional Integration in the Western Brazilian Amazon." *Agroforestry Systems* 38(1–3):195–222.

Whiteman, Adrian, Christopher Brown, and Gary Q. Bull. 1999. "Forest Product Market Developments: The Outlook for Forest Products Markets to 2010 and the Implications for Improving Management of the Global Forest Estate." Working Paper: FAO/FPIRS/02. Food and Agriculture Organization, Forest Policy and Planning Division, Rome, Italy.

World Bank. 1978. "Forestry." Sector Policy Paper. Washington, D.C.

———. 1991a. "Environmental Assessment Sourcebook. Vol. 1: Policies, Procedures, and Cross-Sectoral Issues." World Bank Technical Paper 139. Washington, D.C.

———. 1991b. "Environmental Assessment Sourcebook. Vol. 2: Sectoral Guidelines." World Bank Technical Paper 140. Washington, D.C.

———. 1991c. "Environmental Assessment Sourcebook. Vol. 3: Guidelines for Environmental Assessment of Energy and Industry Projects." World Bank Technical Paper 154. Washington, D.C.

———. 1991d. "Forestry: The World Bank's Experience." Operations Evaluation Department. Washington, D.C.

———. 1991e. "The Forest Sector." A World Bank Policy Paper. Washington, D.C.

———. 1992a. "Brazil: An Analysis of Environmental Problems in the Amazon." OED Report No. 9104-BR. Washington, D.C.

———. 1992b. "Guidelines for Monitoring and Evaluation of GEF Biodiversity Projects." Environment Department, Washington, D.C.

———. 1992c. "Strategy for Forest Sector Development in Asia." World Bank Technical Paper 182. Washington, D.C.

———. 1992d. *World Development Report 1993: Development and the Environment.* Washington, D.C.

———. 1993a. "The World Bank Operational Manual, Operational Policies: Environmental Action Plans." Report Number OP4.02. Washington, D.C.

———. 1993b. "The World Bank Operational Manual, Operational Directives: Environmental Assessment." Report Number OD4.01. Washington, D.C.

———. 1993c. "The World Bank Operational Manual, Operational Directives: Environmental Policies." Report Number OD4.00. Washington, D.C.

———. 1993d. "The World Bank Operational Manual, Operational Policies: Forestry." Report Number OP4.36. Washington, D.C.

———. 1993e. "The World Bank Operational Manual, Operational Directives: Indigenous People." Report Number OD4.20. Washington, D.C.

———. 1993f. "The World Bank Operational Manual, Operational Directives: Involuntary Resettlement." Report Number OD4.30. Washington, D.C.

———. 1993g. "The World Bank Operational Manual, Operational Policies: Natural Habitats." Report Number OP4.04. Washington, D.C.

———. 1993h. "The World Bank Operational Manual, Operational Directives: Poverty Reduction." Report Number OD4.15. Washington, D.C.

———. 1993i. *World Development Report 1993: Investing in Health.* New York: Oxford University Press.

———. 1994a. "Conditional Lending Experience in World Bank-Financed Forestry Projects." OED (Operations Evaluation Department) Report 13820. Washington, D.C.

———. 1994b. "Review of Implementation of the Forest Sector Policy." Agriculture Department, Washington, D.C.

———. 1997. "What Drives Deforestation in the Brazilian Amazon? Evidence from Satellite and Socioeconomic Data." Environment Working Paper. Report WPS1772. Washington, D.C.

———. 1998a. "Approach Paper: Forests and the World Bank. An Operations Evaluation Department Study." Washington, D.C.

———. 1998b. "Forest Policy Implementation Review and Strategy: Initiating Memorandum." Environmentally and Socially Sustainable Development Network, Washington, D.C.

———. 1998c. "Sustaining Tropical Forests: Can We Do It, Is It Worth Doing?" Tropical Forests Expert Discussion Meeting, Meeting Report, Part I. Forests Team, Natural Resources Program, Environment Department, Washington, D.C. Unpublished.

———. 1998d. "Sustaining Tropical Forests: Can We Do It, Is It Worth Doing?" Tropical Forests Expert Discussion Meeting, Meeting Report, Part II. Background Papers. Forests Team, Natural Resources Program, Environment Department, Washington, D.C. Unpublished.

———. 1998e. "Biodiversity in World Bank Projects: A Portfolio Review." Environment Department Working Paper 59. Washington, D.C.

———. 2000a. "Africa Region Environment Strategy." Draft. Africa Region, Washington, D.C.

———. 2000b. *Cleaner Transport Fuels for Cleaner Air in Central Asia and the Caucasus.* Washington, D.C.

———. 2000c. "East Asia and the Pacific: Regional Environment Strategy Summary." Draft. East Asia and the Pacific Region, Washington, D.C.

———. 2000d. "Europe and Central Asia Environment Strategy." Draft. Europe and Central Asia Region, Washington, D.C.

———. 2000e. "Latin America and the Caribbean: Regional Environmental Strategy." Draft. Latin America and the Caribbean Region. Washington, D.C.

———. 2000f. "Natural Resource Management Strategy: Eastern Europe and Central Asia." World Bank Technical Paper 485. Washington, D.C.

———. 2000g. "A Review of the World Bank's Forest Strategy and Its Implementation." Operations Evaluation Department. World Bank, Washington, D.C.

———. 2000h. "Rural Development, Natural Resources and the Environment: Lessons of Experience in Eastern Europe and Central Asia." Eastern Europe and Central Asia, Washington, D.C.

———. 2000i. "South Asia Environment Strategy." Draft. South Asia Region, Washington, D.C.

———. 2000j. "Supporting the Web of Life: The World Bank and Biodiversity. A Portfolio Update (1988–99)." Environment Department, Washington, D.C.

———. 2000k. *The Quality of Growth.* Washington, D.C.

———. 2000l. "The World Bank and the Global Environment: A Progress Report." Environment Department, Washington, D.C.

———. 2000m. *World Development Report 2000/2001: Attacking Poverty.* New York: Oxford University Press.

———. 2000n. "Implementation of the 1991 Forest Strategy in IFC's Projects." International Finance Corporation, Operations Evaluation Group, Washington, D.C.

———. 2001a. "Annual Review of Development Effectiveness: From Strategy to Results." Operations Evaluation Department. Washington, D.C.

———. 2001b. "China: Environment Sector Strategy Update." Environment and Social Development Sector Unit, Washington, D.C.

———. 2001c. "Financing Sustainable Forest Management: Doing More with (Probably) Less." Paper presented by a World Bank representative at the United Nations Forest Forum workshop, Oslo, February 2001.

———. 2001d. "Indonesia: Environment and Natural Resource Management in a Time of Transition." Environment and Social Development Sector Unit, Washington, D.C.

———. 2001e. "Middle East and North Africa Region: Environment Strategy Update." Draft. Middle East and North Africa Region, Washington, D.C.

———. 2001f. "Project Appraisal Document, Papua New Guinea Forestry and Conservation Project." Report 20641-PG. Rural Development and Natural Resources Sector Unit, East Asia and Pacific Region, Washington, D.C.

———. 2002. "Integrating Gender into the World Bank's Work: A Strategy for Action." World Bank, Washington, D.C.

———. Various years. *World Development Indicators [1997–2000].* Washington, D.C.

World Bank/WWF Alliance for Forest Conservation and Sustainable Use. 1998. "Reference Book." July 30. Steering Committee Meeting binder. Washington, D.C.

———. 1999a. *Annual Report 1999.* Washington, D.C.

———. 1999b. "Conversion of Paper Parks to Effective Management: Developing a Target." Paper presented at the International Workshop on Management Effectiveness of Protected Areas, June 14–16, 1999, CATIE, Turrialba, Costa Rica.

World Rainforest Movement, Forest Peoples Programme, and Environmental Defense Fund. 1999. Briefing Paper: "The World Bank's Forest Policy Implementation Review and Strategy Development." www.wrm.org.uy/english/tropical_forests/wb.html.

World Resources Institute. 1988. *Breathing Easier: Taking Action on Climate Change, Air Pollution, and Energy Insecurity.* Washington, D.C.

WWF (World Wide Fund for Nature) International and IUCN (The World Conservation Union). 1999. "Management Effectiveness in Forest Protected Areas." Gland, Switzerland.

INDEX

Note: Page numbers with the prefix *A–* are in the Appendixes, found on the CD-ROM that accompanies this book.

A

B

C

D

E

F

G

H

I

J

K

L

M

N

O

P

R

S

T

U